Chris Defonseka
Flexible Polyurethane Foams

Also of interest

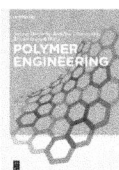

Polymer Engineering
Tylkowski, Wieszczycka, Jastrzab (Eds.), 2017
ISBN 978-3-11-046828-1, e-ISBN 978-3-11-046974-5

Water-Blown Cellular Polymers
A Practical Guide
Defonseka, 2019
ISBN 978-3-11-063950-6, e-ISBN 978-3-11-064312-1

Polymeric Composites with Rice Hulls
An Introduction
Defonseka, 2019
ISBN 978-3-11-063968-1, e-ISBN 978-3-11-064320-6

Two-Component Polyurethane Systems
Innovative Processing Methods
Defonseka, 2019
ISBN 978-3-11-063957-5, e-ISBN 978-3-11-064316-9

e-Polymers.
Editor-in-Chief: Seema Agarwal
ISSN 2197-4586
e-ISSN 1618-7229

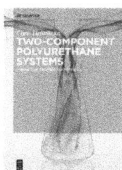

Chris Defonseka

Flexible Polyurethane Foams

A Practical Guide

DE GRUYTER

Author
Chris Defonseka
Toronto
Canada
defonsekachris@rogers.com

ISBN 978-3-11-063958-2
e-ISBN (PDF) 978-3-11-064318-3
e-ISBN (EPUB) 978-3-11-063983-4

Library of Congress Control Number: 2018964973

Bibliographic information published by the Deutsche Nationalbibliothek
The Deutsche Nationalbibliothek lists this publication in the Deutsche Nationalbibliografie;
detailed bibliographic data are available on the Internet at http://dnb.dnb.de.

© 2019 Walter de Gruyter GmbH, Berlin/Boston
Typesetting: Integra Software Services Pvt. Ltd.
Printing and binding: CPI books GmbH, Leck
Cover image: MirageC/gettyimages

www.degruyter.com

Preface

Polyurethanes are an important branch of plastics which have a key role in everyday life. There are many types and grades of polyurethanes, and this book will be about flexible and viscoelastic grades. These foams have become very popular and essential materials in most areas of human activity, especially in industries concerned with comfort.

The use of flexible polyurethane foams have seen rapid growth over the years propelled by constant research and have virtually replaced traditional materials. Versatility, durability and special properties have made polyurethane foams special materials with uses spreading from domestic to industrial applications and even to space travel.

The introduction of viscoelastic ('memory') foam in recent years has enhanced this sector and dramatically changed people's lifestyles. These specially structured polymers provide superior comfort and amazing therapeutic properties that are valuable in the medical field. Continuing research is enabling chemists and chemical manufacturers to produce the basic raw material from sources other than crude oil.

Although polyurethanes are based on complex chemistry (especially polymer chemistry), this book presents the information for ease of understanding and processing methods. The market for flexible polyurethane foam in any part of the world is so large that there is plenty of scope for even very small producers.

Manufacturers of these flexible foams can be classified into large-volume producers needing very large capital outlay, medium producers with single-block moulding, small producers with low-cost start-ups and fabricators from foam block purchases.

It is hoped that the readers of this book will benefit from the in-depth knowledge and information provided by the author, who is active as a consultant in local and international scenes. He brings to this book very valuable experience from rendering assistance to producers (small and large) in several countries. This book provides a thorough understanding of flexible polyurethane foams and practical approaches for manufacturing methodology for the entrepreneur and larger producers from a practical perspective.

https://doi.org/10.1515/9783110643183-201

Acknowledgements

Due to the highly technical nature of this book and in order to provide useful information for the reader, the author gratefully acknowledges the sources stated below. He also thanks the Commissioning Editor and the Editorial Degruyters for their excellent support and guidance. The author offers practical information based on hands-on and international experience, which he hopes will be greatly beneficial to entrepreneurs and for the producers of flexible polyurethane foam already in business:

1. Foam machinery photographs courtesy of Modern Enterprises, India.
   http://www.foam-machinery.com
2. Foam machinery photographs courtesy of AS Enterprises, India.
   http://www.as-enterprises.com
3. Foaming Fundamentals by Modern Enterprises Limited, India.
4. Low-Pressure Metering Equipment – an article by the American Chemistry Council.
5. Continuous Foaming Plants for Flexible Slab Stock – an article by the Canon Group.
6. Flexible and Viscoelastic Polyurethane Foams by Chemcontrol Limited.
7. Glossary of Flexible Polyurethane Foam Technology by the Polyurethane Foam Association.
8. Moulded Foams – Dow Chemical Corporation – for client information.
9. Evaluation and Testing – Dow Chemical Corporation – for client information.
10. Urethane Foam Properties – Technical Leaflet by Foam-Tech.
11. A Processing Guide for Integral Skin Polyurethanes.
12. Troubleshooting Guide for Flexible Polyurethane Foams – processing guide by Dow Chemical Corporation.
13. Bio-Polyols – a technical leaflet by BioBased Technologies.
14. Glossary of Flexible Polyurethane Foam Terms – Publication by the Edge Sweets Company.

https://doi.org/10.1515/9783110643183-202

Contents

1 Introduction

1.1 Brief history

Polyurethane (PUR) foams were first introduced into the market in the 1950s. They are commonly known as 'polyurethane foams', or 'PU foams'. With rapid development and growth over the years, these are available under three main categories – (1) flexible, (2) semi-rigid and (3) rigid – of which the most popular one is the flexible foam.

1.1.1 Flexible foams

This new material gained great interest and common acceptance initially in the mattress sector as a superior replacement for traditional materials such as rubber, fibre and cotton because of its unique properties. Compared with rubber latex foam, flexible PUR foams are superior in many ways. They can be produced in densities ranging from 12 to 50 kg/m³. Flexible PUR foams have maximum tear strength, better resistance to oxidation, aging, as well as fire, and its open cell structure enables it to absorb and emit humidity (e.g. body heat).

In the latter years, a new PUR foam with special properties has been developed and introduced into the market. This is viscoelastic foam and is commonly called 'memory foam'. This foam soon became very popular because of its medicinal properties, in addition to the superb comfort it provided and some even called it 'miracle foam'. The viscoelastic foam was initially too expensive for widespread use; however, over the years, it became cheaper and more affordable.

1.1.2 Rigid polyurethane foams

Rigid PUR foam is a key insulating material used in residential and commercial construction industries, such as for spray PUR foam roofing, PUR laminated panels, insulated sheathing and single-skin and foamed panels for garage doors. These foams are also used as binder applications such as particle board, laminated stranded lumber, moisture-resistant rim board and moisture-resistant flooring. Additionally, there are many speciality applications such as wheels and core binders.

https://doi.org/10.1515/9783110643183-001

1.1.3 Automotives

Automotive applications include interior and exterior parts. Some are energy-absorbing PUR foams, dashboard panels, steering wheels, interior and exterior trims, head liners and many others.

1.1.4 Elastomers

For customers who are looking for an elastic, durable and wear-resistant material for demanding applications, the PUR cast elastomers are ideal. These systems are used to produce rollers, wheels, vibration-damping parts and encapsulants. Spray PUR elastomer systems (known as 'thick-film coatings') coat and protect a wide range of wood, metal and concrete surfaces.

1.1.5 Insulation

Many versatile energy-efficient PUR systems are available to the appliance industry, especially the makers of household refrigerators and commercial refrigeration systems, as well as water-heater manufacturers.

1.1.6 Moulded parts

For applications of design flexibility, low production volumes and product durability, these PUR systems can be used in reaction injection moulding for a wide variety of non-automotive parts and components. From agricultural equipment body panels, to the snowboard cores to the trim for heavy-duty trucks, rigid PUR foams, elastomeric PURs and solid PUR can be an ideal solution to produce lightweight, durable moulded parts. Newer technologies are also expanding the use of the PUR system-I composites.

1.1.7 Spray roofing systems

Spray PUR foam roofing systems, high-performance coatings, sprayed insulation and adhesives offer a range of commercial construction roofing solutions to contractors and building owners. These products provide durable, trouble-free performance in any climate and under any weather condition.

1.1.8 Speciality urethane systems

Diphenylmethane based PUR and elastomer systems are available for filters, foot-wear, integral skin foams, carpet underlay, turf backing, flexible moulded foams and rigid foams.

These are some of the main categories of PUR. Flexible foams and viscoelastic foams are the most versatile and utilised foams, and have the widest range of applications.

This incredible material initially dominated the domestic market with end uses such as mattresses, pillows, cushions and mattress toppers. It has also spread into other sectors such as industrial and automotive applications, medical applications and space travel. The basic raw material components for flexible PUR are *polyols*. Polyols are petroleum-based compounds and, in keeping with global environmental concerns, continuous research has produced polyols from soya and other sources, making these foams more cost-effective.

1.2 Types of foams

Many types of foams are on the market. The popular ones are: polystyrene, poly-ethylene, polyester-based PUR and polyether-based PUR. All these foams are syn-thetic foams belonging to the plastic family, with two very desirable properties: (1) being easily malleable or moulded, and (ii) capable of 'softening' and regaining its original shape (especially flexible PUR foams). Where comfort is concerned, poly-ether-based PUR foams give the best results because of their extra-soft texture and they also cost less than polyester-based foams (which are rough and give a 'scratchy' surface).

Polyether PUR foams have wide applications in furniture, bedding, pillows, padding and carpet underlay. Polyester-based PUR foams are used for textiles, shoulder pads, noise reduction and other applications. Both are used singly or in combination for applications in automotive, aircraft, household and footwear industries. These foams are made from base chemicals (polyols) derived from liquefied ethylene gas (which is obtained during the refining of crude oil). Polyols called 'ecopolyols' that are based on vegetable oils and other sources have been developed. With rapid development programmes, these new polyols may replace the traditional ones quite soon.

During recent years, mattresses in the bedding sector have been made from materials such as coir, cotton, latex foam and flexible PUR foams. The use of steel springs and coils are being fast replaced with PUR foams to maximise comfort. Flexible PUR foams are very important in our day-to-day life. The discovery and introduction of viscoelastic foam have escalated the demand for PUR flexible foams and, according to reports, it is a billion-dollar market in North America alone.

In the 1960s, Swedish scientists, at the request of NASA, developed a special PUR foam to counter the G-forces their astronauts experienced during take-off. They produced a new type of PUR foam called viscoelastic foam. This incredible foam material has achieved a very high profile in the bedding market (especially in the mattress sector) because of its unique properties of proven maximum comfort and medical benefits.

Viscoelastic foams have very low resilience and damping effects. They continually adjust to the shape of the human body and its weight, thereby dramatically reducing pressure points, creating a sensation of 'floating on a cloud' and ensuring maximum comfort with improved blood circulation, guaranteeing a tossing-free, deep sleep. This material has been thoroughly tested commercially and medically by researchers, and is highly recommended as the best 'sleep' material on the market. Unlike other foams, viscoelastic foams are four dimensional in that they have density, hardness, temperature and time effects, properties that make them truly special foams.

1.3 Basic foam grades

Cheap grades are foams where quality is not a priority. These include soft or coarse foams used for gymnasium mats or camping sleep pads and carpet underlay (where the foams are heavily loaded with fillers). Cheaper base polymers called 'graft' polymers are also used to reduce costs. Foam blocks made from re-bonded foam scrap and cut to desired sizes are used for carpet underlay and thick slabs as bases for mattresses.

Medium soft grades are good-quality foams used for comfort (e.g. back cushions on a couch or a window seat). Here, the density and firmness of the foam are important factors. They are also used for low-end inexpensive foam mattresses.

*Firm quality grades*are foams used for high-end or special applications such as bedding, aircraft seats, furniture and auto seats. Here, the important properties are firmness, the 'spring effect', anti-fungal, ultraviolet (UV)-protected, fire-retardant and any other property as desired by a customer or on market demand.

High-end grades are called 'high-resiliency (HR) foams' and are used for special applications depending on the end use or customer preference. These foams can have some or all of the properties mentioned above.

Viscoelastic foam is a quality foam, like no other. It is used for luxury, comfort and special applications such as medical uses and space travel. It is a four-dimensional foam that is much heavier and costlier than standard PUR foams, with densities ranging up to $\geq 96 \, \text{kg/m}^3$. It is used as luxury mattresses with thicknesses ranging from 4 inches to 8 inches. This speciality foam is also used in pillows, pads and many other applications.

1.4 Viscoelastic foam mattresses: Marketing hype or molecular miracle?

In the last decade, a new type of flexible foam burst onto the foam market and came to be known as a viscoelastic or 'slow' foam. It quickly became popular and was also called memory foam because of its property of full recovery.

Its immediate impact and marketing led some people to believe that this was a 'magic' material because it had quite amazing and unusual properties. In a comparatively short time, it was apparent that the claims of its 'magical' properties were not marketing 'hype' by the manufacturers. The marketplace was excited by the fact that it was suitable for 'space travel' (which is a story in itself) but also had many benefits. Foam-market experts realised that anybody can promote a well-made product, but that a different situation arose if the product itself had major advantages over competing materials for comfort and health features with proven therapeutic values.

A key feature of viscoelastic foam is that it is 'lazy' or slow with high damping. One of the main characteristics is very low resilience, meaning that there is no 'spring' in the foam. Another important aspect is the temperature sensitivity of the foam (which is a key active property of viscoelastic foams). If one were to 'sink' a hand into viscoelastic foam, it leaves an indentation that disappears as the foam recovers slowly. Unlike conventional flexible foams, which are two dimensional (density and hardness can be varied), viscoelastic foams are four dimensional (hardness, density, temperature and time can be varied). Most flexible foams are highly resilient and 'bouncy', whereas a viscoelastic foam allows moulding to the body shape. The low indentation force deflection (IFD) of these foams allows a body to sink rather deep into the foam without resistance while maintaining the 'firm' feel of good-quality foam.

Because of their special medical benefits, many hospitals are now using viscoelastic foam mattresses for their beds. Conventional flexible foams (especially HR foams) cause pressure points on a body but viscoelastic foams, under the influence of body heat, soften and allow the body to sink in, while firmly cushioning it. The soft foam flows around the pressure points, thus spreading the load over a larger area and creating negative pressure. For patients who have to lie in bed for prolonged periods, this property prevents the creation of ulcers ('bed sores'). Another beneficial aspect of viscoelastic foam is the promotion of free-flowing blood circulation for a body lying prone for long periods, such as during surgery. This foam cushions the body without pressure, so there are many benefits for people with pain in the back or joints, and this has created a market demand.

Viscoelastic foam mattresses have been replacing conventional mattresses for some time and, although these mattresses are more expensive, people have been willing to pay the extra cost, which they think is justified as compared with the benefits of using one. Another way to avoid paying high prices is to use a 2″ or 3″ 'mattress topper' on top of the conventional mattress, which is very effective. There are other products of viscoelastic foam products now available on the market: pads, pillows, footwear and medical cushions.

1.5 Flexible slabstock foams

There are four main types of slabstock foams: (1) conventional, (2) high-resilience, (3) filled foam and (4) high load-bearing. Each type is formulated differently to provide varying characteristics depending on the intended market.

Conventional slabstock –Standard formulations can be modified readily to produce a wide range of foam properties. These foams can be broadly classified into four basic general hardness or softness grades. In general, the hardness of a foam is defined by the IFD, which can be manipulated by changing or increasing/decreasing some of the components of a foam formulation.

As a general guide, foams having a density ≈ 16 kg/m^3 and an IFD of 10–20 are 'super soft', whereas foams with densities of 32–45 kg/m^3 and an IFD>40 are considered to be 'hard'. Anything in between can be considered to be 'soft' to 'intermediate'.

HR foams are made in a wide range of densities and hardness levels, and are designed primarily to offer enhanced support. These foams are more 'elastic' and feel heavier than standard foams. They are more expensive than other slabstock foams, so their uses are usually limited to high-performance products.

Filled foams – Filled slabstock foams use inorganic fillers to increase the foam density and also to improve the load-bearing characteristics. These foams often lead to the perception that they are higher-quality products. Filled foams decrease quality by reducing the tensile strength, elongation and tear strength, and perhaps causing a long-term decrease in the fatigue resistance of the foam. In general, calcium carbonate is used as the filler but melamine fillers can also be used for improved flammability resistance in applications in which stringent flammability standards are required, such as aircraft seats, automobile seats and public occupancy products.

High load-bearing foams incorporate a graft polyol in the formulation to increase the hardness of the foam. However, manufacturers must carefully balance the enhancement in hardness against density increase to avoid decreasing other physical properties of the foam. Otherwise, high load-bearing foams may suffer from reduced fatigue resistance. High load-bearing foams are commonly used for the production of carpet underlay.

Graft polyol-filled foams are also useful for reducing pressure conditions in high-altitude products and speciality applications such as packaging.

Grade identification – A single manufacturer is likely to make more than one grade of flexible PUR foam and in varying densities. To identify these different grades, a manufacturer will use different colours and codes. Re-bonded foam hardly needs an identification code because the appearance itself – multi-coloured cores – is sufficient.

All end uses for flexible PUR foams (except maybe for sponges and a few others) need finishing (e.g. mattress and cushion covers, upholstery for seats). In general, the materials used are cotton or more expensive fabrics, artificial leather made from foamed polyvinyl chloride, coated cloth with PUR coatings and quilting.

1.6 What is a flexible polyurethane foam?

A flexible PUR foam is a soft, open cell and pliable material belonging to the *thermosetting group* of plastics, and is made by the combination of several chemicals. The main ingredients that make up a basic PUR foam are: polyols as the base, toluene diisocyanate or MDI, water and methylene chloride as blowing agents, amine catalysts and tin catalysts.

The incorporation of additives will give the foam special properties such as colour, UV protection, high resiliency, fire retardation and anti-fungal protection. These foams can be made in different densities and qualities to meet the needs of final applications by adjusting the components of formulations. Polyols can be based on polyethers or polyesters and, although both will produce open cell foams, the latter (due their smaller cell structure) have tighter air flow and are less flexible. Polyether-based foams are spongier and softer and the preferred material, if comfort is the prime consideration.

Ecopolyols (e.g. from soya) will also produce open cells but may give lower yields. Foams made from all three types of polyols are widely used in very large volumes in many applications all over the world.

Viscoelastic foams are also based on polyols and can be classified under flexible PUR foam. They are produced (more or less) by the same mixture of chemicals except that different special grades of polyols are used.

1.7 What is a viscoelastic foam?

A viscoelastic foam is a flexible PUR foam specially formulated to increase its density and viscosity, and has low-resilience and pressure-sensitive properties. Its four-dimensional properties allow maximum comfort by quickly moulding itself to any body shape or weight. When the body is removed, the foam quickly regains its original shape.

Viscoelastic foam has an open cell structure that reacts to body heat and weight and relieves pressure by zero resistance. Depending on the formulations and end applications, cell structures can vary from very open to almost closed cells. The tighter the cell structure, the lower is the airflow through the foam. Quality 'breathable' viscoelastic foam has a good open cell structure, allowing high airflow, good recovery and low odour retention. Most viscoelastic foam products (especially mattresses) are vacuum-packed because of their heavy weight. They may require a short time after removal from packing to remove the slight odour 'build-up'.

Some researchers feel that, instead of the IFD rating, the density of a viscoelastic foam is a more important measure of quality, but this may not be true. In general, the market for these foam mattresses with respect to quality is based on the density and IFD rating, which vary with different types of products. Foam densities of $\geq 80 \, \text{kg/m}^3$ are considered to be of high quality, whereas a density of $64-72 \, \text{kg/m}^3$ is fairly common; densities lower than $64-72 \, \text{kg/m}^3$ are seldom used for mattresses. A good, generally accepted standard for viscoelastic foam mattresses has a density of $\approx 72-80 \, \text{kg/m}^3$, and for mattress toppers slightly lower densities are acceptable but the minimum thickness should be $\geq 2''$ for comfort.

A viscoelastic foam mattress is denser than other foam mattresses, making it more supportive but also heavier. Naturally, these foam mattresses and other products are more expensive than other similar products. Comfort in a mattress is generally measured by its firmness, which indicates its softness or hardness. Firmness is measured by the IFD rating. IFD measures the force required to make a dent of $1''$ into a foam sample $15'' \times 15'' \times 4''$ by a disc having an $8''$ diameter. This is known as IFD @ 25% compression. IFD ratings for viscoelastic foams range between IFD 10 (super soft) and IFD 120 (semi-rigid). Most viscoelastic foam mattresses are from IFD 12 to IFD 16.

Viscoelastic foams work by conforming to the natural heat, weight and shape of a body. They adjust to accommodate the various pressure points such as the shoulder and hips, which are the areas that press hardest into a bed. Depending on which position someone sleeps in, pressure points differ. However, without any support, any sleep position can result in pain or discomfort. Viscoelastic foam with tight but open cells passes air to adjacent cells when pressed down. In this way, even if the mattress or pillow feels firm at first, it can quickly mould to the shape of a body or head and cushion them.

Research and practical use have shown the ability of viscoelastic foam to provide maximum comfort better than other traditional materials, especially its ability to achieve a better night sleep. Viscoelastic foams are less likely to harbour dust mites, which tend to breed, reproduce and proliferate in beds using traditional materials for mattresses. Dust mites result in allergic reactions and other respiratory problems, making them very undesirable pests. The chances of this happening in viscoelastic foams are very remote because of their structure and high densities.

1.8 Characteristics of viscoelastic foams

Viscoelastic foams are typified by their slow recovery after compression. If a weighted object (e.g. a human body) is positioned on a viscoelastic foam, it will progressively conform to the shape of the body and, if it is removed, the foam will slowly recover to its original state. Because of this gradual recovery, viscoelastic foam can also be described as 'slow recovery' foam. Other characteristics include the ability of viscoelastic foams to dampen vibration and absorb shock. Standard ball rebound tests done in a laboratory have confirmed a rebound of <20% as compared with other varieties of flexible PUR foams with a rebound of 50–60%, supporting the view that viscoelastic foams can be termed as 'dead' or 'low-resilience' foams lacking the 'springiness' of other flexible foams. The lack of resilience may appear to be a disadvantage, but a dead foam can be highly desirable in some applications. In addition to these key advantages, many viscoelastic products also react to body temperature and ambient temperatures, softening with heat and conforming to any shape and size of body contour, thereby providing maximum comfort and giving the sensation of 'floating on a cloud'.

1.9 Physical properties of viscoelastic foams

Density – As with all flexible PUR foams, the density of viscoelastic foams are also calculated and expressed in weight/volume units – kg/m^3 or g/cm^3. With conventional flexible polyurethane foams (FPF), a higher foam density generally equates to increased foam durability (retention of performance properties). The same is true of viscoelastic foams. Density enhances the durability and ability of viscoelastic foams to maintain their physical performance. Viscoelastic foam products found in most household and healthcare product applications may typically range in density from $48 \, kg/m^3$ to $96 \, kg/ \, kg/m^3$.

Firmness – Normally, the firmness of viscoelastic foams can range from super soft to semi-rigid. The potential for reduction in surface pressure is closely associated with firmness and can vary based on the formulation. Viscoelastic products with a lower IFD tend to exhibit increased conformance and can distribute body weight more efficiently to alleviate pressure. However, if the IFD is very low and there is not sufficient density or thickness to provide support, the product may 'bottom out', negating the benefits of pressure reduction. When conducting IFD tests, firmness measurements of viscoelastic foams can be significantly affected by some of the key characteristics of the foam: rate sensitivity (the rate of recovery after compression of the foam), sensitivity to temperature and sensitivity to humidity. Sample conditioning before testing is important.

Rate sensitivity – The rate sensitivity of viscoelastic foams, observed as the speed that a foam sample recovers after compression, affects the way firmness (IFD) can be determined in laboratory tests. Because of rate sensitivity, when testing a viscoelastic foam under load (as in IFD measurement), the speed at which the weight force is applied can alter firmness readings. That is, if the indentation plate used in the test descends quickly, the foam may respond with stiffness, whereas slowerspeeds may result in different IFD measurements. With viscoelastic products, IFD tests should show notation of the process speed (rate of deflection) being used so that valid and fair comparisons among foam grades can be made.

Sensitivity to temperature – The physical properties of viscoelastic foams can be influenced greatly by temperature. Even slight changes in room temperature can affect measured firmness and recovery rates. Recovery rate has been positively correlated to heat, so that as the foam increases in temperature, pliability and compression and recovery rates increase. In colder conditions, viscoelastic products tend to become firmer or even stiff. Depending on the formulation, some viscoelastic products can maintain their memory feature at as low as −1 °C but the optimum range for best memory action is typically between 26 and 40 °C. Research is going on to find ways to moderate pliability and allow viscoelastic foams to perform within a broader spectrum of temperatures (especially below freezing). This is especially important in vehicle seating and other applications, where ambient temperatures cannot be controlled in use.

In testing the performance characteristics of viscoelastic foams, making note of the ambient temperature and being certain that all types of comparison testing undertaken under identical conditions are important. Samples must be conditioned before testing. Heat (even body heat) can soften viscoelastic materials and, in extreme situations, affect the ability of the foam to provide support. This process is known as a 'phase change of relaxation'. Rather than broadening the temperature range at which viscoelastic foams retain their slow recovery, the foam formulation can be adjusted to narrow the temperature span at which phase change occurs, anticipating the influence of room temperature combined with body temperature. That is, if the phase change occurs at lower temperatures, its reaction to added body heat could lessen its firmness to the point where support is lost. If the phase relaxation were to occur in warmer conditions, added body heat would not be sufficient to soften the foam, and the resulting foam rigidity would reduce the potential for pressure relief. Hence, careful control of the characteristics of phase change is crucial for the foam to serve its pressure-reduction purpose and to provide predictable support.

Sensitivity to humidity – Viscoelastic foams react to humidity as well as temperature. Viscoelastic products tend to soften in more humid conditions. For example, a very pliant foam may feel 'slick' or 'buttery' as opposed to coarse depending on humidity. Surface feel is generally not a critical factor in end-use applications because (in almost all cases) the foam is enclosed within a covering material, which provides its own surface feel. Firmness that changes with ambient conditions can be difficult for accurate calculation of physical properties.

1.10 Performance testing of viscoelastic foams

1.10.1 Reduction and relief of pressure

Because viscoelastic foam can closely conform to the shape of the human body, it can efficiently distribute pressure over the whole surface. Pressure-mapping equipment is often used to calculate the degree of weight distribution. During the mapping procedure, computer programmes monitor pressure. The body image is graphed to show which areas endure the most pressure (namely the shoulder blades, posterior region, head, heels, calves and elbows). Some producers of viscoelastic foam carry out these tests as an indication of how well the foam might act to minimise pressure. In the healthcare area, to be labelled as providing 'pressure relief', readings must be at ≤32 mmHg, whereas 'pressure reduction' performance is agreed to occur between 32 and 50 mmHg. If readings made on viscoelastic products are compared with those of conventional foam and other common cushioning materials, the pressure differences are highly notable.

Some manufacturers offer viscoelastic foam at ≈85 kg/m^3, whereas other products may range from 40 to 60 kg/m^3 depending on which factory it is made. Viscoelastic foams are very expensive, 3–4 times the price of conventional foams; however, well-known manufacturers and brand names have no difficulty in marketing them because of the obvious and proven superior comfort and health benefits. Other than the most popular uses in bedding and pillows, viscoelastic foams are used in prostheses and medical devices. For example, a layer of viscoelastic foam can be used under plaster of Paris for setting broken bones and in neck braces, as well as in trolley pads in accident and emergency units in hospitals. Wheelchairs (in which patients may stay for long periods) can use three layers of this special foam – (1) soft, (2) firm and a (3) lot firmer – to maximise comfort and to ensure there is no bottoming out. Viscoelastic foam is now being used for seating but, because it is expensive, may not be suitable for the mass markets and suit only high-end products because people are willing to pay these prices. A relatively new area of use is for moulded viscoelastic foam parts for office furniture and similar uses. Another interesting area of application is in seating for orchestras, where the musicians have to sit at different angles to play their instruments, and special energy-absorbing viscoelastic foams fulfil his need.

1.10.2 Hyper or pseudo effects?

When viscoelastic foam came on the market, there were many debates as to what 'viscoelastic' meant. In simple terms, whereas conventional flexible foams are made from standard polyols, viscoelastic foams are made from viscoelastic special polymers. While manipulating cell structures to leave vestigial cell walls to create similar initial behaviour, these foams are not truly viscoelastic because good viscoelastic foams need an open cell structure. Foams with a lot of closed cells can seem slow because, if cells are crushed, it takes time to recover, but this is just slowness in taking up air again. Some inferior imitations of viscoelastic foam are marketed and it is easy to detect these foams because it is difficult to draw air out of them. A simple test to determine the true quality of a viscoelastic foam is to put a piece of foam in a freezer for 1 h and it will be very hard (rigid), whereas an inferior quality foam will only be soft or semi-rigid. Conventional or HR foams have been used in hospital beds. Viscoelastic foams have slowly replaced them with the great additional benefit of stopping bed sores and providing pressure relief as well as maximum comfort for patients and are of special value for long-term patients.

1.10.3 Viscoelastic foam: summary

Viscoelastic foam technology represents new and exciting developments for the flexible PUR foam industry. It has taken hold in various markets (from bedding to medical to technical applications) and offers a unique cushioning alternative to conventional and HR flexible PUR foams. Its performance in specialised end-user applications has been well received by consumers. Ongoing efforts in industry and individuals are focused on improving the control of the performance of viscoelastic foam so that more versatile and successful applications can be developed for this unique cushioning, pressure-distributing and shock-absorbing material:
- Viscoelastic foam (also known as memory foam) is a type of open-cell, flexible PUR foam.
- Surface comfort and pressure distribution are closely related to the ability of the foam to conform to the shape of the body.
- Slow recovery adds comfort but also complicates the testing procedures.
- The firmness, support and height recovery rate of viscoelastic foam are often affected by temperature and body heat.
- Comparative laboratory testing must match ambient testing conditions with regard to accuracy.
- Viscoelastic products should be thoroughly ventilated after production and during fabrication to reduce the possibility of bothersome aromatic emissions.
- Some standardised FPF performance tests may require special consideration to accommodate the slow recovery rate of viscoelastic foam.

1.10.4 End applications

All flexible PUR foams are very versatile materials and are used singly or in combination with everyday applications such as mattresses, cushions, sheets, slabs, pillows, quilting, padding, sound proofing, toppers, sponges, carpet underlay, rebonded foam, aircraft seats, sponge cloths, textiles, footwear, furniture, automobile seats, medical needs, gym mats, utility mats, apparel, camp pads, beach pads, egg crate toppers, convoluted pads, futons, industrial sponges, packaging, wheelchair pads, cervical pillows, space travel and day care furniture.

2 Basic chemistry

2.1 Brief introduction

Chemistry is the study of the composition of matter and how that composition changes. The two broad areas of chemistry are organic and inorganic. Organic chemistry deals with matter that contains the element carbon, whereas inorganic chemistry is the study of matter mineral in origin. The term 'organic' originally used to mean compounds of plants or animals, but now it also includes many synthetic materials that have been developed through research. One such group of synthetic organic materials is called 'plastics', under which urethanes are classified.

Chemical theory states that all matter consists of simple base substances called 'elements' such as hydrogen, oxygen and nitrogen. Matter is defined as anything that has mass (weight) and occupies space, and may exist in the form of a solid, liquid or gas. The structures of matter are composed of small particles called 'atoms', which combine to form larger particles called 'molecules' and, when these combine, they form compounds. The smallest particle size is an atom, whereas a molecule is formed by two or more atoms. Atoms consist of particles called 'neutrons', 'protons' and 'electrons'. The atoms of each element differ in the number of these particles they contain. Thus, no two elements will have identical atoms. The neutron is a particle with no electrical charge, whereas protons are positively charged. Both occupy the centre of an atom called the nucleus. Electrons are negatively charged particles and are in orbit around the nucleus. The negative charge of the electrons moving around the nucleus is equal to the positive charge of the protons in the nucleus. Therefore, for all elements, an atom is electrically neutral. If an atom gains an electron, it becomes electronegative (a negative ion) and if an atom loses an electron, it becomes electropositive (a positive ion). An ion is an atom that has gained or lost one or more electrons.

The atomic number of an element is the number of protons in the nucleus of an atom. The atomic weight of an element is the mass of an atom of that element compared with the mass of an atom of carbon. The molecular mass or the molecular weight is the total atomic weights of the atoms making up the molecules. The Periodic Table lists all 118 elements and also shows their atomic and molecular weights. When single elements join together they form compounds.

For example, hydrogen + oxygen = water
sodium + chlorine = sodium chloride = common salt
hydrogen + nitrogen = ammonia

Some of the most common elements used in combination with basic plastic units are hydrogen, oxygen, nitrogen, chlorine, fluorine, silicon and carbon. They form many polymers, including urethanes, and thus polyurethanes (PURs).

https://doi.org/10.1515/9783110643183-002

2.2 Organic chemistry

2.2.1 Basics

Organic chemistry is a sub-discipline within chemistry involving the study of the structure, properties, composition, reactions and preparation by synthesis (or other means) of carbon-based compounds, hydrocarbons and their derivatives. These compounds may contain any number of other elements, including hydrogen, nitrogen, oxygen, halogens as well as phosphorus, silicon and sulfur.

Organic compounds are structurally diverse. The range of applications of organic compounds is enormous. They form the basis of or constituents of many products: plastics, drugs, food, petrochemicals, explosives and paints. They form the basis of almost all earthly life processes.

2.2.2 History

Before the nineteenth century, chemists, in general, believed that compounds obtained from living organisms were too complex to be synthesised. They believed that such organic matter was endowed with a 'vital force'. They named these compounds 'organic'.

During the first half of the nineteenth century, scientists realised that organic compounds can be synthesised in the laboratory. The study of soaps from various fats and alkalis enabled them to separate the different acids that, in combination with alkalis, produced soap. These were all individual compounds, so it was possible to make a chemical change in the fats (which came from traditional organic sources) to produce new compounds. The crucial breakthrough in organic chemistry was the concept of chemical structure, which was developed independently by two scientists. Both men suggested that tetravalent carbon atoms could link to each other to form a carbon lattice, and that the detailed patterns of atomic bonding could be discerned by skilful interpretations of appropriate chemical reactions.

The surge of organic chemistry continued with the discovery of petroleum and its separations into fractions. The conversion of different compound types or individual compounds by various chemical processes created the petroleum industry. This led to the vast petrochemical industry, which successfully manufactured artificial rubber, various organic adhesives, property-modifying additives and versatile plastics. Another major industry – pharmaceuticals – began in the last decade of the nineteenth century when the manufacture of acetylsalicylic acid (commonly known as aspirin) was made by a famous chemical company in Germany.

Although early examples of organic reactions and applications were often by 'trial and error', the latter half of the century witnessed highly systematic and well-organised studies of organic compounds. Beginning in the early twentieth century, progress in organic chemistry allowed the synthesis of highly complex molecules using multi-step procedures. Concurrently, polymers and enzymes were understood to be large organic molecules, and petroleum was found to be of biological origin. Finding new methods of synthesis for a given compound is called 'total synthesis'. Total synthesis of complex natural compounds started with urea followed by glucose in 1907, and camphor was used for total synthesis for commercial purposes. Since then, substantial benefits have been afforded pharmaceutical manufacturers, and a booming and expanding market has been the result.

2.2.3 Characteristics

Organic compounds often exist as mixtures, so various methods have also been developed to assess purity, such as chromatographic methods, Traditional methods of separation include distillation, crystallisation and solvent extraction.

Organic compounds were traditionally characterised by various chemical tests called 'wet methods' but such tests have been largely replaced by more modern methods such as spectroscopic or other computer-related methods of analyses. There are four main analytical methods:

- Nuclear magnetic resonance (NMR) spectroscopy is probably the most commonly used method, and often permits complete assignment of atom 'connectivity' and even stereochemistry using correlation spectroscopy. The principal constituent atoms of organic chemistry, hydrogen and carbon, exist naturally with NMR.
- An elemental analysis is a destructive method used to determine the elemental composition of a molecule.
- Mass spectrometry indicates the molecular weight of a compound and, from the fragmentation patterns, its structure. High-resolution mass spectrometry can usually be used to identify the exact formula of a compound and is in lieu of elemental analyses. Previously, mass spectrometry was restricted to neutral molecules that exhibited some volatility but advanced methods allow us to obtain the mass spectrometry of any organic compound.
- Crystallography is an unambiguous method for determining molecular geometry, the proviso being that single crystals of the material must be available, and that the crystal must be representative of the main sample. Highly automated software allows a structure to be determined within a short period of time. Traditional spectroscopic methods such as infrared spectroscopy, optical rotation and ultraviolet spectroscopy provide relatively non-specific information but remain in use for specific classes of compounds.

2.2.4 Properties

Physical properties of typical organic compounds of interest include quantitative and qualitative features. Quantitative information includes melting point, boiling point and index of refraction, whereas qualitative properties include odour, colour, consistency and solubility.

2.2.5 Melting and boiling properties

In contrast to many inorganic compounds, organic materials melt and boil. The melting point and boiling point provide crucial information about the purity and identity of organic compounds. This information correlates with the polarity of molecules and their molecular weights. Some organic compounds (especially symmetrical ones) undergo sublimation (transition from the solid phase to the gas phase without passing through an intermediate liquid phase). A well-known example of a sublime organic compound is p-dichlorobenzene, the odiferous constituent of modern-day mothballs. Organic compounds are usually not very stable at temperatures ≥300 °C (although there are exceptions).

2.2.6 Solubility

Neutral organic compounds tend to become hydrophobic (i.e. they are less soluble in water than in organic solvents). Exceptions include organic compounds that contain ionisable groups as well as low molecular weight alcohols, amines and carboxylic acids, where hydrogen bonding occurs. Organic compounds tend to dissolve in organic solvents. Solvents can be pure substances such as ether or mixtures. Mixtures can be paraffinic solvents such as the various petroleum or tar fractions obtained by physical separation or by chemical conversion. The degree of solubility in different solvents is dependent upon the solvent type and on the functional groups (if present).

2.2.7 Nomenclature

The names of organic compounds can be systematic (following logically from a set of rules) or non-systematic (following various traditions). Systematic nomenclature starts with the name of the parent structure within the molecule of interest. This parent name is then modified by prefixes, suffixes and numbers to convey the structure unambiguously. Given that millions of organic compounds are known, rigorous use of systematic names can be cumbersome. Thus, these rules are followed closely for simple compounds but not for complex molecules. To use the systematic naming procedure, one must know the structure and names of the parent structure. Parent structures include unsubstituted hydrocarbons, heterocycles and monofunctionalised derivatives thereof.

Non-systematic nomenclature is simpler and unambiguous (at least to chemists). Non-systematic names do not indicate the structure of a compound. Non-systematic names are common for complex molecules (which includes most natural products). Computer technology has enabled other naming methods to evolve that are intended to be interpreted by machines.

2.2.8 Structural presentation

Organic molecules can be described by drawings, structural formulae, combinations of drawings and chemical symbols. The line-angle formula is simple and unambiguous. In this system, the endpoints and intersections of each line represent one carbon atom, and hydrogen atoms can be notated explicitly or assumed to be present as implied by a tetravalent carbon. The depiction of organic compounds with drawings is greatly simplified by the fact that carbon in almost all organic compounds has four bonds, oxygen has two bonds, hydrogen has one bond and nitrogen has three bonds.

2.3 Classification of organic compounds

2.3.1 Functional groups

The concept of functional groups is central to organic chemistry as a means to classify structures and for predicting properties. A functional group is a molecular model and the reactivity of that functional group is assumed (within limitations) to be the same in various molecules. Functional groups can have a decisive influence on the chemical and physical properties of organic compounds. Molecules are classified on the basis of their functional groups.

For example, alcohols all have the subunit C–O–H. All alcohols tend to be hydrophilic (usually from esters) and can be converted to the corresponding halides. Most functional groups feature heteroatoms (atoms other than carbon and hydrogen). Organic compounds are classified according to their functional groups, alcohols, carboxylic acids and amines.

2.3.2 Aliphatic compounds

Aliphatic hydrocarbons are subdivided into three groups of homologous series according to their state of saturation:

- Paraffin – which are alkanes without any double or triple bonds
- Olefins or alkenes – which contain one or more double bonds (e.g. polyolefins)
- Alkynes – which have one or more triple bonds

The rest of the groups are classified according to the functional groups present. Such compounds can be straight-chained, branch-chained or cyclic. The degree of branching affects characteristics such as the octane number as in the petroleum industry.

Saturated (alicyclic) compounds and unsaturated compounds exist as cyclic derivatives. The most stable rings contain five or six carbon atoms, but large rings, such as macro-cycles and smaller rings, are also common. The smallest cycloalkane family is the three-membered cyclopropane. Saturated cyclic compounds contain single bonds only, whereas aromatic rings have alternating or conjugated double bonds. Cycloalkanes do not contain multiple bonds, whereas cycloalkenes and cycloalkynes do.

2.3.3 Aromatic compounds

Aromatic hydrocarbons contain conjugated double bonds. The most important example is benzene, and the structure of which was formulated by a chemist who first proposed delocalisation (also known as the resonance principle) to explain its structure. For conventional cyclic compounds, aromaticity is conferred by delocalised electrons. Particular instability (anti-aromaticity) is conferred by conjugated electrons.

2.3.4 Heterocyclic compounds

The characteristics of cyclic hydrocarbons are altered again if heteroatoms are present. These can exist as substituents attached to the ring or as a member of the ring itself. In the latter, the ring is termed a 'heterocycle'. Pyridine and furan are the corresponding alicyclic heterocycles. The heteroatom heterocyclic molecules are, in general, oxygen, sulfur or nitrogen, with the latter being particularly common in biochemical systems.

Examples of groups among heterocyclic compounds are aniline dyes; alkaloids, vitamins, nucleic acids [e.g. deoxyribonucleic acid (DNA), ribonucleic acid]; and numerous medicines. Rings can fuse with other rings on an 'edge' to give polycyclic compounds. Rings can also fuse on a 'corner' such that one atom (usually carbon) has two bonds going to one ring and two to another. Such compounds are termed 'spiro' and are important in several natural products.

2.3.5 Polymers

An important property of carbon is that it forms chains or networks that are linked by carbon–carbon bonds. This linking process is called 'polymerisation', whereas the chains or networks are called 'polymers'. The starting source compounds are called 'monomers', the basic unit of a plastic.

There are two main groups of polymers: synthetic polymers and biopolymers. Synthetic polymers are manufactured artificially and commonly referred to as 'industrial polymers'. Biopolymers occur within natural environments or with human intervention. Since the invention of the first synthetic polymer product – Bakelite – many synthetic polymers have been invented. Common synthetic organic polymers are polyethylene (commonly called 'polythene'), polypropylene, nylon, Teflon, polystyrene, polymethylmethacrylate (called 'Plexiglas' and 'perspex') and polyvinyl chloride. Synthetic and natural rubbers are also polymers.

Varieties of each synthetic polymer product may exist for a specific use. Changing the conditions at the time of polymerisation alters the chemical structure of the final product and its properties. These changes include chain lengths, branching or tacticity. If the product is from a single polymer, it is called a 'homo-polymer'. Secondary components may be added to create a 'hetero-polymer' (co-polymer) and the degree of clustering of the different components can also be controlled. Physical properties such as hardness, density, mechanical or tensile strength, abrasion, heat resistance, transparency and colour are dependent upon the composition of additives in relation to the end use.

2.3.6 Organic synthesis

Synthetic organic chemistry is an applied science because it borders engineering design, analyses and/or construction of works for practical purposes. The organic synthesis of a simple compound is a problem-solving task in which a synthesis is designed for a specific molecule by selecting optimal starting materials. Complex compounds can have tens of reaction steps that 'build' the desired molecule sequentially. The synthesis proceeds by utilising the reactivity of the functional group in the molecule. For example, a carbonyl compound can be used as a nucleophile by converting it into an electrophile. The combination of the two is called the 'aldol reaction'. Designing practically useful syntheses involves conducting the synthesis in the laboratory. The practice of creating synthetic routes for complex molecules is called total synthesis.

There are several strategies in designing a synthesis. A method splicing of the target molecule to pieces based on known reactions. The pieces (or the proposed precursors) receive identical treatment until available (and ideally inexpensive) starting materials are reached. Then, the 'retro-synthesis' is written in the opposite direction to give the synthesis. A 'synthetic tree' can be constructed because each compound and each precursor has multiple syntheses.

2.3.7 Organic reactions

Organic reactions are chemical reactions involving organic compounds. Pure hydrocarbons undergo limited classes of reactions, but many more reactions which organic compounds undergo are determined primarily by the functional groups. The general theory of these reactions involves careful analyses of the electron affinity of the key atoms, bond strengths and steric hindrance. These issues can determine the relative stability of short-lived reactive intermediates, which usually directly determine the path of the reaction.

The basic reaction types are addition, elimination, substitution, pericyclic, rearrangement and redox. The number of possible organic reactions is infinite. However, certain general patterns are observed that can be used to describe the many common and useful reactions. Each reaction has a stepwise reaction mechanism that explains how it happens in sequence (even though a detailed description of steps is not always clear from a list of reactions alone).

2.4 Inorganic chemistry

Inorganic chemistry is the branch of chemistry dealing with the properties and behaviour of non-carbon compounds. The distinction between organic chemistry and inorganic chemistry is far from absolute and there is much overlap, mostly in the sub-discipline of organometallic chemistry.

Many inorganic compounds are ionic compounds consisting of cations and anions joined by ionic bonding. Examples of ionic compounds are salts such as magnesium chloride, which consists of magnesium cations and chlorine anions. Another example is sodium oxide, which consists of sodium cations and oxide anions. In any salt, the proportions of ions are such that the electric charges cancel out, so that the bulk compound is electrically neutral. Ions are described by their oxidation state. Their ease of formation can be inferred by the ionisation potential (cations) or from the electron affinity (anions) of the sourcing elements.

Important classes of inorganic salts are oxides, carbonates, sulfates and halides. Many inorganic salts are characterised by high melting points. Inorganic salts are typically poor conductors in the solid state. Another important feature is their solubility in water and ease of crystallization. Some salts such as sodium chloride are very soluble in water, but others are not. The simplest inorganic reaction is double displacement: in the mixing of two salts, the ions are swapped without a change in oxidation state. In redox reactions, a reactant, the oxidant, lowers its oxidation state and another reactant has its oxidation state increased. The net result is an exchange of electrons. Electron exchange can also occur indirectly, for example in batteries, which is a key concept in electrochemistry.

If a reactant contains hydrogen atoms, then a reaction can take place by exchanging protons in acid-based chemistry. In a more general definition, an acid can be any chemical species capable of binding to electron pairs. Conversely, any molecule that tends to donate an electron pair is referred to as a 'base'. As a refinement of acid–base interactions, the polarisability and size of ions should also be taken into account.

Inorganic compounds are found as minerals in nature. Soil may contain iron sulfide as pyrite or calcium sulfate as gypsum. Inorganic compounds are also found 'multitasking' as biomolecules such as electrolytes (sodium chloride), in energy storage or in construction (polyphosphate, the backbone of DNA). The first important man-made inorganic compound was ammonium nitrate for soil fertilisation. Inorganic compounds are synthesised for use as catalysts such as vanadium and titanium chloride or as reagents in organic chemistry such as lithium aluminium hydride. Subdivisions of inorganic chemistry are organometallic chemistry, cluster chemistry and bio-inorganic chemistry. These fields are active areas of research in inorganic chemistry aimed at finding new catalysts, superconductors and therapies.

2.4.1 Industrial inorganic chemistry

Traditionally, inorganic chemistry is a major sector of production of many countries and a major force in enhancing their economies. Some of the most popular and essential inorganic chemicals manufactured are ammonia, aluminium sulfate, ammonium nitrate, carbon black, chlorine, hydrochloric acid, hydrogen, hydrogen peroxide, nitric acid, nitrogen, oxygen, phosphoric acid, sodium carbonate, sodium chloride, sodium hydroxide, sodium silicate, sodium sulfate, sulfuric acid and titanium dioxide. The manufacture of fertilisers is another major sector of industrial inorganic chemistry.

2.5 Classification of inorganic chemistry

Partly, the classification focuses on the position in the periodic table of the heaviest element (the element with the highest atomic weight) in a compound, and partly by grouping compounds by their structural similarities. When studying inorganic compounds, one often encounters the different classes of inorganic chemistry. An organometallic compound is characterised by its coordination chemistry and may show interesting solid-state properties.

2.5.1 Coordination compounds

Classical coordination compounds feature metal bonds to 'lone pairs' of electrons residing in the main group atoms of ligands such as water and ammonia. In modern coordination compounds, almost all organic and inorganic compounds can be used as ligands. From a certain perspective, all chemical compounds can be described as 'coordination complexes'.

2.5.2 Main group compounds

Main group compounds have been known since the beginning of chemistry (e.g. elemental sulfur and distillable white phosphorus). Experiments in oxygen by chemists identified not only an important diatomic gas but opened the way for describing compounds and reactions according to stoichiometric ratios. The discovery of a practical synthesis of ammonia using ion catalysts demonstrated the importance of inorganic chemical analyses. Typical main group compounds are silicon dioxide, tin chloride and nitrous oxide. Many main group compounds can also be classified as 'organometallic' because they contain organic groups. Main group compounds also occur in nature (e.g. phosphate in DNA) and therefore may be categorised as 'bio-inorganic'. Conversely, organic compounds lacking many hydrogen ligands can be classified as 'inorganic', such as fullerenes and binary carbon oxides.

2.5.3 Transition metal compounds

Compounds containing metals from group 4 to group 11 of the periodic table are considered to be transition metal compounds. Compounds with a metal from group 3 to group 12 are sometimes also incorporated into this group but are often also classified as main group compounds. Transition metal compounds show rich coordination chemistry, varying from tetrahedral for titanium to square planar for some nickel complexes to octahedral for coordination complexes of cobalt. A range of transition metals can be found in biologically important compounds, such as iron in haemoglobin.

2.5.4 Organometallic compounds

Organometallic compounds are considered to be a special category because organic ligands are often sensitive to hydrolysis or oxidation. Thus, organometallic chemistry employs more specialised preparative methods than traditional chemistry. Synthetic methodology (especially manipulating complexes in solvents of low coordinating power) has enabled exploration of very weak coordinating ligands such as hydrocarbons. Because the ligands are petrochemical in some sense, organometallic chemistry is very relevant to industry.

2.5.5 Cluster compounds

Clusters can be found in all classes of chemical compounds. According to the commonly accepted definition, a cluster consists of a triangular set of atoms that are bonded directly to each other. Metal–metal-bonded dimetallic complexes are highly relevant in this area. Clusters occur in 'pure' inorganic systems, organometallic chemistry, main group chemistry and bio-inorganic chemistry. The distinction between very large clusters and bulk solids is increasingly blurred. This interface is the chemical basis of nanotechnology and arises specifically from the study of quantum size effects in cadmium-selenide clusters. Thus, large clusters can be described as an array of bound atoms intermediate between a molecule and a solid.

2.5.6 Bio-inorganic compounds

By definition, bio-inorganic compounds occur in nature but the subclass includes anthropogenic series such as pollutants and drugs. This research area incorporates many aspects of biochemistry, including many types of compounds and metal complexes containing ligands that range from biological macromolecules to the ill-defined humic acid. Traditionally, bio-inorganic chemistry focuses on electron transfer and energy transfer in proteins relevant to respiration. Medicinal inorganic chemistry includes the study of essential and non-essential elements with application to diagnoses and therapies.

2.5.7 Solid state compounds

This important area focuses on the structure, bonding and physical properties of materials. In practice, solid state inorganic chemistry uses methods such as crystallography to gain an understanding of properties that result from collective interactions between the subunits of the solid. Included in solid state chemistry are metals and their alloys or intermediate derivatives. Related fields are condensed physics, mineralogy and material science.

2.5.8 Qualitative theories

Inorganic chemistry has benefited greatly from qualitative theories. Such theories are easier to learn because they require little understanding of quantum theory. Within main group compounds, a popular theory predicts (or at least rationalises) the structures of main group compounds, such as an explanation why some are pyramidal and others are T-shaped. For transition metals, the crystal field theory allows one to understand the magnetism of many simple complexes. A particularly powerful qualitative approach for assessing structure and reactivity begins with classifying molecules according to electron counting by focusing on the numbers of valence electrons (usually at the central atom in a molecule).

2.5.9 Molecular symmetry group theory

A central construct in inorganic chemistry is the theory of molecular symmetry. Mathematical group theory provides the language to describe the shapes of molecules according to their point group symmetry. Group theory also enables factoring and simplification of theoretical calculations. As an instructional tool, group theory highlights commonalities and differences in the bonding of otherwise disparate species such as carbon dioxide and nitrogen dioxide.

2.6 Basics of analytical chemistry

Analytical chemistry is the study of the separation, identification and quantification of chemical compounds of natural and artificial materials. Qualitative analyses give an indication of the identity of the chemical species in the selected sample, and quantitative analyses determine the amount of one or more of these components. The separation of components is often undertaken before analyses.

Analytical methods can be separated into 'classical' and 'instrumental'. Classical methods (also known as 'wet chemistry methods') use separations such as precipitation, extraction, distillation and qualitative analyses by colour, odour or melting point. Quantitative analyses are achieved by measurement of weight or volume. Instrumental methods use an apparatus to measure the physical quantities of the analyte such as light absorption, fluorescence or conductivity. The separation of materials is accomplished using chromatography or electrophoresis methods. Analytical chemistry has applications in forensics, bio-analyses, clinical analyses, environmental analyses, materials analyses and is important for industrial chemists.

2.6.1 Qualitative analyses

A qualitative analysis determines the presence or absence of a particular compound but not the mass or concentration. That is, it is not related to quantity.

2.6.1.1 Flame test

An inorganic qualitative analysis refers to a systematic method to confirm the presence of certain (usually aqueous) ions or elements by carrying out a series of reactions that eliminate range of possibilities and then confirm the suspected ions with a test. Sometimes, small carbon-containing ions are included in such methods. With modern instrumentation these tests are rarely used but can be useful for educational purposes and in situations where access to state-of-the-art instruments is not available or expedient.

2.6.1.2 Gravimetric analyses

Gravimetric analyses involve determining the amount of material present by weighing the selected sample before and after transformation. A common example used for educational purposes is the determination of the amount of water in a hydrate by heating the sample to remove the water such that the difference in weight is due to the loss of water.

2.6.1.3 Volumetric analyses

Titration involves the addition of a reactant to a solution being analysed until an equivalence point is reached. Often the amount of material in the solution being analysed may be determined. Most familiar is an acid–base titration involving a colour-changing indicator. There are many other types of titrations, for example, potentiometric titrations. These may use different types of indicators to reach an equivalence point.

2.6.1.4 Mass spectrometry

Spectrometry measures the interaction of molecules with electromagnetic radiation. Spectrometry consists of many different applications such as atomic absorption, atomic emission, NMR and a host of others.

2.6.1.5 Electrochemical analyses

Electroanalytical methods measure the potential (volts) and/or current (amperes) in an electrochemical cell containing the analyte. These methods can be categorised according to which aspect of the cell is being controlled and measured. The three main categories are potentiometry (difference in electrode potentials), coulometry (cell current measured over time) and voltammetry (cell current is measured while actively altering the cell potential).

2.6.1.6 Microscopy

The visualisation of single molecules, single cells and nanomaterials is an important and attractive approach in analytical science. Microscopy can be categorised into three fields: optical, electron and scanning. These fields have seen rapid growth and development because of developments in computer science and camera industries.

All these branches of chemistry and continuing research are very important and form the basis for polymer chemistry, which leads to the formation and manufacture of plastic compounds, of which PURs are a vital branch.

3 Basic polymer chemistry

Polymer chemistry is a very complex subject. The basic information presented in this chapter is sufficient for the reader to gain a good understanding of polymers, which forms the base for plastics under which polyurethanes (PURs) are classified.

In ancient times, people used copper, gold, iron and clay for their daily needs, which were found in their natural form in the earth. However, as time passed and technology advanced, people needed to make goods that were strong, malleable, durable, practical and cheap. Materials with these properties could not be found with what was available, so they looked for newer materials, which led to synthetic materials. One such group of materials is plastics.

The word 'plastics' is derived from the Greek word *plastikos*. In general, a plastic material can be defined as a material that is pliable and capable of being shaped by temperature and pressure. Plastics are based on polymers, derived from the Greek word *polymeros* with *poly* meaning 'many' and *meros* meaning 'basic units'. Polymers are also called 'resins', which is not really correct because resins are gum-like substances.

3.1 What is a polymer?

A polymer is a large molecule ('macromolecule') composed of repeating basic structural units. These subunits are typically connected by covalent chemical bonds. Although the term 'polymer' is sometimes taken to refer to plastics, it encompasses a larger class of compounds comprising natural and synthetic materials with a wide variety of properties. Because of the extraordinary range of properties of polymeric materials, they have essential and vital roles in everyday life. This role ranges from the familiar synthetic plastics and elastomers to natural biopolymers such as nucleic acids and proteins, which are essential for life.

Natural polymeric materials such as shellac, amber and natural rubber have been used for centuries. Many other natural polymers also exist such as the well-known cellulose, the main constituent of wood and paper. Examples of synthetic polymers (Figure 3.1) are synthetic rubber, neoprene, nylon, polyvinyl chloride (PVC), polystyrene, polyethylene, polypropylene, silicone and PUR. The most common continuously linked backbone of a polymer used for the preparation of plastic compounds consists mainly of carbon atoms. A simple example is polyethylene (commonly known as 'polythene'), whose repeating unit is based on the ethylene monomer, a petroleum byproduct. However, other elements such as oxygen are also commonly present in polymer backbones. Most elements are combinations of two or more atoms bonded together to form molecules.

https://doi.org/10.1515/9783110643183-003

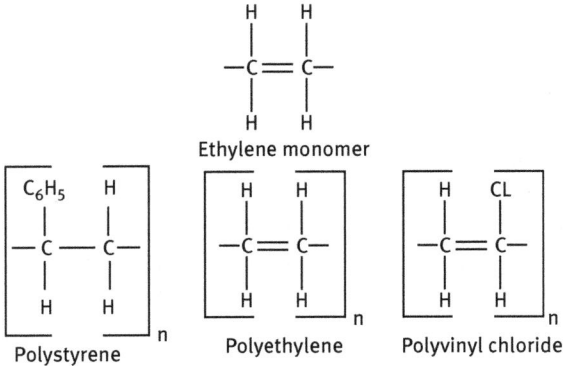

Figure 3.1: Basic polymer structures.

In plastics, bonding patterns are very important because they determine the physical characteristics of the plastic product. Bonds can be divided into primary bonds (ionic or covalent bonds) and secondary bonds (van der Waals forces). All bonds between atoms and elements are electrical in nature. Most elements constantly try to reach a stable state by (a) receiving extra electrons, (b) releasing electrons or (c) sharing electrons.

A polymer is made up of many basic units ('mers') repeated in a pattern and, if a large number of molecules bond together, a polymer is formed. These basic units are considered to be bifunctional with two reactive bonding sites. Monomers (Figure 3.2) are the initial basic units used in forming polymers and, in most cases, they are liquids. The chemical reaction or process that joins these monomers together is called 'polymerisation'. If the polymer consists of similar repeating

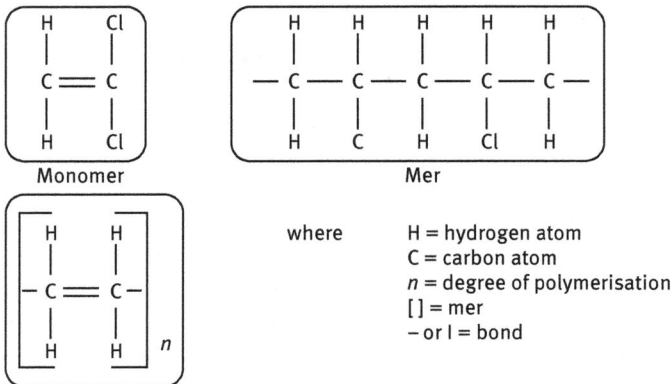

where H = hydrogen atom
       C = carbon atom
       n = degree of polymerisation
       [] = mer
       – or | = bond

Figure 3.2: Basic monomers example.

units, it is called a 'homopolymer' (meaning 'the same'). If two different types of monomers are polymerised, it is called a 'copolymer'. If three types of monomers are polymerised to form a polymer, it is called a 'terpolymer' (Figure 3.3) (these are also known as copolymers). An 'elastomer' is described as any polymer that can be stretched to ≥200% of its original length.

Polymers and copolymers

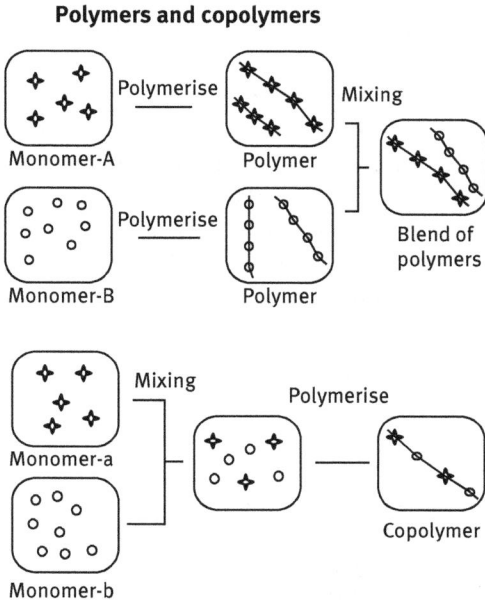

Figure 3.3: Formation of polymers.

For example, M + M + M + · · ·   = homopolymer
                M1 + M2 + · · ·   = copolymer
                M1 + M2 + M3 + · · ·. = terpolymer

In general, polymers are classified by systems based on source, light penetration, heat reaction, polymerisation and crystal structure:
- Source: polymers from natural sources as well as modified natural or synthetic sources
- Light penetration: optical properties (e.g. opaque, transparent or luminescent)
- Heat reactions: thermoplastic (soft on heating, solid on cooling; can be re-used)
- Polymerisation: a method of joining many monomers (basic units) to form polymers and copolymers
- Crystal structure: crystalline (molecules arranged in order) and amorphous (random arrangement of molecules)

3.1.1 Polymer synthesis

Polymerisation is based on combining many small molecules known as monomers (basic units) to covalently bonded chains or networks. During this process, some chemical groups may be lost from each monomer. This is the case, for example, in the polymerisation of polyethylene terephthalate polyester. The monomers are terephthalic acid and ethylene glycol but the repeating unit is $-OC-C_6H_4-COO-CH_2-CH_2-O-$, which corresponds to the combination of the two monomers with the loss of two water molecules. The distinct portion of each monomer that is incorporated into the polymer is known as a 'repeat unit' or 'monomer residue'. Laboratory synthetic methods are generally divided into two categories: step-growth polymerisation and chain-growth polymerisation. The essential difference between the two is that in chain-growth polymerisation, monomers are added to the chain one at a time only, whereas in step-growth polymerisation, chains of monomers may combine with one another directly. Synthetic polymerisation reactions may be carried out with or without catalysts.

3.1.2 Modification of natural polymers

Many commercially important polymers are synthesised by the chemical modification of naturally occurring polymers. Prominent examples include the reaction of nitric acid and cellulose to form nitrocellulose as well as the formation of vulcanised rubber by heating natural rubber in the presence of sulfur. Some of the ways in which polymers can be modified include oxidation, cross-linking and end-capping.

3.2 Polymer properties

Polymer properties can be divided into several classes based on the scale at which the property is defined, as well as upon its physical basis. The most basic property of a polymer is the identity of its constituent monomers. A second set of properties, known as 'microstructure', describes the arrangement of these monomers within the polymer at the scale of a single chain. These basic structural properties play a major part in determining the bulk physical properties of the polymer (which describes how the polymer behaves as a continuous macroscopic material). Chemical properties describe how the chains interact through various physical forces. At the macroscale, they describe how the bulk polymer interacts with other chemicals and solvents.

3.2.1 Monomers

A monomer (basic unit) of a plastic material is the initial molecule used in forming a polymer. If a large number of molecules bond together, a polymer is formed. One of the most important and common monomers is ethylene gas resulting from the refinement of crude oil, which is liquefied. Most monomers are liquids.

Monomers form five basic structures: (i) linear, (ii) cross-linked, (iii) branched, (iv) graft and (v) interpenetrating. If two different monomers are joined in sequence, it is called a 'copolymerisation' and a copolymer plastic material is the result:

1. Linear structures – there are three basic linear copolymer classes:
   - Alternating – **ABABABAB**...
   - Random – **ABBBAAABAABAB**...
   - Block – **AAABBBBBAAA**...
2. Cross-linked structures – these involve primary covalent bonds between the molecular chains:

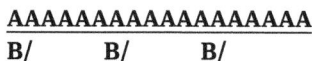

   <u>**AAAAAAAAAAAAAAAAAA**</u>
     B/   **B/**   **B/**

3. Branched structures represent three-dimensional branching of the linear chain.
4. Graft structures – graft polymers are similar to cross-linked and branch-structured copolymers. The term 'graft' has been used to denote a controlled or engineered addition to the main polymer chain.

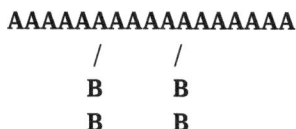

   AAAAAAAAAAAAAAAAA
       /   **/**
      B   **B**
      B   **B**

5. Interpenetrating structures – these polymer networks are an entangled combination of two cross-linked polymers that are not bonded to each other. A 'plastic alloy' is formed if two or more different polymers are physically mixed together in the melt. The term 'polyblend' refers to a plastic that has been modified by the addition of an elastomer.

The identity of the monomer residues (repeating units) forming a polymer is its first and most important attribute. Polymer nomenclature is generally based upon the type of monomer residues comprising the polymer. Polymers that contain only a single type of repeating unit are known as homopolymers, whereas polymers containing a mixture of repeating units are known as copolymers. Polystyrene, for example, is composed of only styrene monomer residues and is therefore classified as a homopolymer. Ethylene vinyl acetate contains more than one variety of monomer and is thus a copolymer.

3.2.2 Microstructure

The microstructure of a polymer (sometimes called 'configuration') relates to the physical arrangement of monomers along the backbone of the chain. These are the elements of polymer structure that require breaking of a covalent bond in order to change. Structure has a strong influence on the other properties of a polymer. For example, two samples of natural rubber may exhibit different durabilities even though their molecules comprise the same monomers.

An important microstructural feature of a polymer is its architecture, which relates to the way branch points lead to a deviation from a simple linear chain. A branched polymer molecule is composed of a main chain with one or more substituent side chains or branches. Types of branched polymers include star polymers, comb polymers, brush polymers, dendronised polymers and dendrimers. The architecture of a polymer affects many of its physical properties, including (but not limited to) solution viscosity, melt viscosity, solubility in various solvents, glass transition temperature (T_g) and the size of individual polymer coils in solution. Various methods may be employed for the synthesis of a polymeric material with a range of architectures.

3.2.3 Chain length

The physical properties of a polymer are strongly dependent on the size or length of the polymer chain. For example, as chain length is increased, melting and boiling temperatures increase. Impact resistance also tends to increase with chain length, as does the viscosity or the resistance to flow of the polymer in its melt state.

A common means of expressing the length of a chain is the degree of polymerisation, which quantifies the number of monomers incorporated into the chain. As with other molecules, the size of the polymer may also be expressed in terms of molecular weight. Synthetic polymerisation methods typically yield a polymer product including a range of molecular weights, so weight is often expressed statistically to describe the distribution of chain lengths present. Common examples are the number average molecular weight and weight average molecular weight. The ratio of these two values is the polydensity index. A final measurement is contour length, which can be understood as the length of the chain backbone in its fully extended state.

3.2.4 Molecular mass (weight)

The mass of a molecule is determined by adding together the atomic masses of the various atoms making up the molecule:

For example, the atomic mass of hydrogen is 1 and that of oxygen is 16.
Thus, the molecular mass of a single molecule of water $(H_2O) = 1 + 1 + 16 = 18$.

The length (molecular mass) of a polymer chain has a profound effect on the processing and properties of a plastic compound. Increasing chain length may increase toughness, melt temperature, melt viscosity, creep and stress resistance. It may also increase the difficulty of processing, material costs and possible polymer degradation.

3.2.5 Polymerisation

The mechanism or chemical reaction that joins the basic monomers is called polymerisation. Polymerisation falls into two general categories: addition and condensation.

Most polymers are produced by addition polymerisation. In this process, unsaturated hydrocarbons are caused to form large molecules. The mechanism of the reaction involves three basic phases:
1. Initiation – the phase in which the monomers are caused to join by the aid of a high-energy source, such as a catalyst or heat source.
2. Propagation – the addition process continues.
3. Termination – the final product. A polymer has been produced and the process is made to stop.

3.2.6 Condensation polymerisation

Condensation is a chemical reaction in which a small molecule (often water) is eliminated from the polymer.

Plastic compounds are formed by the addition of additives such as fillers, solvents, stabilisers, plasticisers and colourants to these basic polymers. These additives influence many characteristics of the final products. Plastics are categorised into two main groups: thermoforming and thermosetting. In the former, a plastic material softens on heating and can be reused, whereas thermosetting materials harden on heating and cannot be reused. It is the molecular arrangement of a polymer that determines the category. Examples of thermoforming plastic resins are polystyrene and polyethylene, whereas examples of thermosetting are PURs and silicones.

3.3 Mechanical properties

The mechanical or bulk properties of a polymer are of primary interest for processing because they dictate how the polymer behaves in use.

3.3.1 Tensile strength

The tensile strength of a material quantifies how much stress the material will endure before suffering permanent deformation. This property is very important in applications that rely upon the physical strength or durability of a polymer. For example, a rubber band with a higher tensile strength will hold a greater weight before breaking. In general, tensile strength increases with the chain length of a polymer and cross-linking of polymer chains.

3.3.2 Young's modulus of elasticity

Young's modulus quantifies the elasticity of a polymer. It is defined as the ratio of the rate of change of stress to strain. Like tensile strength, this is highly relevant in polymer applications involving the physical properties of polymers. The modulus is strongly dependent upon temperature. Viscoelasticity describes a complex time-dependent elastic response when the load is removed. Dynamic mechanical analyses are used to measure the complex modulus by oscillating the load and measuring the resulting strain as a function of time.

3.3.3 Melting point

Melting point, when applied to polymers, suggests not a solid–liquid phase transition but a transition from a crystalline or semi-crystalline phase to a solid amorphous phase. It is more appropriately called the 'crystalline melting temperature'. Among synthetic polymers, crystalline melting is discussed only with regard to thermoplastics because thermosetting polymers decompose at high temperatures rather than melting.

3.3.4 Glass transition temperature

A parameter of particular interest in the manufacture of synthetic polymers is the T_g. This is the temperature at which amorphous polymers undergo a transition from a rubbery, viscous amorphous liquid to a brittle, glassy amorphous solid. T_g can be engineered by altering the degree of branching or cross-linking in a polymer or by the addition of plasticisers. Inclusion of plasticisers tends to lower the T_g and increase the flexibility of the polymer. Plasticisers are generally small molecules that are chemically similar to the polymer and create gaps between polymer chains for greater mobility and reduced inter-chain interactions. A good example of the action of plasticisers is related to PVC. Unplasticised PVC is used for pipe manufacture because it needs to be strong and resistant to heat. Plasticised PVC is used for clothing, artificial leather and sheeting, applications for which flexibility is needed.

3.3.5 Mixing behaviour

In general, polymeric mixtures are far less miscible than mixtures of small molecule materials. This effect results from the fact that the driving force for mixing is usually entropy, not interaction energy. Miscible materials usually form a solution not because their interaction with each other is more favourable than their self-interaction but because of an increase in entropy and hence free energy associated with increasing the amount of volume available to each component. Polymeric molecules are much larger and generally have much higher specific volumes than small molecules, so the number of molecules involved in a polymeric mixture is far smaller than the number in a small molecule mixture of equal volume. The energetics of mixing are comparable on a per volume basis for polymeric and small molecule mixtures.

Furthermore, the phase behaviour of polymer solutions and mixtures is more complex than that of small molecule mixtures. Whereas most small molecule solutions exhibit only an upper critical solution temperature phase transition at which phase separation occurs with cooling, and polymer mixtures commonly exhibit a lower critical solution temperature phase transition at which phase separation occurs with heating.

In dilute solution, the properties of the polymer are characterised by the interaction between the solvent and the polymer. In a good solvent, the polymer appears swollen and occupies a large volume. In this case, intermolecular forces between the solvent and monomer subunits dominate over intermolecular interactions. In a bad or poor solvent, intermolecular forces dominate and the chain contracts.

3.3.6 Polymer degradation

Polymer degradation is a change in the properties – tensile strength, colour, shape or molecular weight – of a polymer or polymer-based product under the influence of one or more environmental factors, such as heat, light and chemicals. It is often due to a change in polymer chain bonds via hydrolysis, leading to a decrease in the molecular mass of the polymer. Although such changes are undesirable in some instances, such as biodegradation and recycling, they may be intended to prevent environmental pollution. Degradation can also be useful in biomedical applications. For example, a copolymer of polylactic acid and polyglycolic acid is employed in hydrolysable stitches that slowly degrade after they are applied to a wound.

The susceptibility of a polymer to degrade is dependent upon its structure. Epoxies and chains containing aromatic functionalities are especially susceptible to degradation by ultraviolet light, whereas polyesters are susceptible to degradation by hydrolysis, and polymers containing an unsaturated backbone are especially susceptible to ozone cracking. Carbon-based polymers are more susceptible to thermal degradation than inorganic polymers and are therefore not ideal for most high-temperature applications.

4 Polyurethane raw materials

The basic raw materials for flexible polyurethane (PUR) foams are polyols, isocyanates, water, methylene chloride, amine catalysts, tin catalysts, chain extenders, colourants, air and additives [e.g. fire retardants, antioxidants, ultraviolet (UV) protectors]. These raw materials are combined on 'systems' rather than a straightforward formula, and are formulated to achieve predetermined properties to suit end applications. Some of the important properties are the final density of the foam, cell structure (texture) and the support factor indentation force deflection. All formulae or systems using these raw materials are designed around these vital aspects.

4.1 Polyols

Most flexible PUR foams are made with polyether polyols. These are long-chain alcohols that are made by polymerising common hydrocarbon oxides. This results in linkages that connect the hydrocarbon portions of the chain and hydroxyl functional groups at the end of the chain. Triols with molecular weights from 3,000 to 4,000, made by copolymerising a mixture of propylene oxide and ethylene oxide with glycerin as the initiator, are often used to produce conventional flexible slabstock foams. Water and traces of alcohol act as co-initiators to produce some diols and monols, so the overall functionality of the polyol is between 2 and 3. Polyols with slightly higher functionalities are also available commercially.

Several properties of foams can be controlled by varying the functionality of a polyol. For example, increasing the polyol functionality without changing the molecular weight produces a slight increase in foam hardness and a small reduction in tensile strength, tear strength and elongation. In addition, as functionality increases, the time at which the gel point occurs decreases. Increasing the equivalent weight of the polyol (molecular weight divided by the functionality) while maintaining the functionality of a polyol produces a foam with increased tensile strength and elongation. However, the increased equivalent weight also reduces the reactivity of the polyol.

Altering the level and distribution of ethylene oxide in the molecular chain can have a dramatic effect on the reactivity, emulsifying capacity and affinity for water of the polyether. For example, a high level of ethylene oxide in the recipe will increase the reactivity of the polyol. If the ethylene oxide is added stepwise rather than as a mixture of oxides, a block polyol is formed rather than a random polyol. If the resulting ethylene oxide block is at the end of the polyol chain, reactivity will be increased even further because the terminal hydroxyl groups now form a more reactive primary alcohol. As the level of ethylene oxide within the polyol increases, the hydrophilicity (affinity for water) and emulsifying capacity also increase.

https://doi.org/10.1515/9783110643183-004

Polyols are liquids and are available commercially in drums, totes or tankers for very-large-volume manufacturers. Their shelf-life is approximately 6 months, and for optimum yield they should be kept at room temperature ≤25 °C and should be stirred before use. During storage time, precautions should be taken to prevent water absorption, especially from the atmosphere.

4.1.1 Graft polyols

Graft polyether polyols contain co-polymerised styrene and acronitriles. They are designed to maximise load-bearing properties in PUR foams. The latest grades of graft polyols have solids in the range of 10–45% in the production of slabstock foam.

4.1.2 Isocyanates

The most common isocyanate used in the manufacture of flexible foams is toluene diisocyanate (TDI). Although diphenylmethane diisocyanate and its polymeric forms can also be used, mixtures of 2,4-TDI and 2,6-TDI are the isocyanates of choice for the production of flexible PUR foams. TDI is a low-cost, high-quality product that allows manufacturers to produce many types of flexible foams with a wide range of physical properties. Although TDI is available in 2,4-TDI:2,6-TDI isomer mixtures of 80:20 or 65:35, as well as the pure 2,4-TDI isomer, the reactivity of the system and the resulting foam properties can be modified using blends of the various isomer-ratio mixtures.

4.1.3 Bio-polyols

Until recently, most polyols originated from petroleum byproducts. The need for alternate sources (especially due to global environmental issues) led to research that resulted in bio-polyols. Bio-polyols are polyols derived from vegetable oils such as soya, canola and peanut oil, and are made by different methods. These bio-based materials are used primarily for the production of PURs.

Bio-polyols may have similar sources and applications but the materials themselves can be quite different depending on how they are made. All are clear liquids, ranging from colourless to slight yellow. Their viscosities also vary, and are usually a function of the molecular weight and average number of hydroxyl groups per molecule, with higher molecular weights and higher hydroxyl content giving higher viscosities. Odour is a significant property, which is different from polyol to polyol, but most bio-polyols are quite similar chemically to the original parent vegetable oil and are prone to becoming rancid. This involves autoxidation of fatty acid chains containing carbon–carbon double bonds and the ultimate formation of odoriferous, low-molecular-weight aldehydes, ketones and carbolic acids. Odour is undesirable in the bio-polyols itself, but it is more important that this aspect is neutralised in the foam materials made from them.

Of the number of naturally occurring vegetable oils that contain the unreacted hydroxyl groups that account for the important reactivity of these polyols, castor oil may be the only commercially available natural oil polyol that is produced directly from a plant source. Others require chemical modification of the oils directly available from plants.

Vegetable oils such as soybean oil, peanut oil and canola oil contain carbon–carbon double bonds but not hydroxyl groups. Several processes are used to introduce hydroxyl groups into the carbon chain of fatty acids and most of them involve oxidation of the carbon–carbon double bond. Castor oil has found numerous applications, many of them due to the hydroxyl groups that allow chemical derivatisation of the oil. Castor oil undergoes most of the reactions that alcohols do but the most important one is the reaction with diisocyanates to make PUR. Castor oil has been used for making various PUR products, ranging from coatings to foams as well as the use of derivatives of castor oil to be an area of active research and development. Castor oil derivatives with propylene oxide makes flexible PUR foams suitable for mattresses (see Chapter 1).

Apart from castor oil, which is a relatively expensive vegetable oil and is not produced domestically in many industrialised countries, the use of polyols derived from vegetable oils to make PUR products has attracted attention from the manufacturers of PUR products, especially producers of flexible foam slabstock. The rising costs of petrochemical feedstock and an enhanced global desire for environmentally friendly 'green' products have created a demand for these materials. These bio-based polyols have a good future because some of the large car manufacturers have started using flexible foams made with them.

The goal is that using renewable resources as feedstock for global chemical processes will reduce the 'environmental footprint' by reducing the demand for non-renewable fossil fuels currently used in the chemical industry and thereby reduce the overall production of carbon dioxide, the most notable 'greenhouse' gas.

Isocyanates are liquids and are available in drums, totes or tankers for large-volume manufacturers. The shelf-life is approximately 6 months. Due to their high flammability and corrosive properties, suitable precautions should be taken at all times. Even at the time of transfer from the tanker to the holding tanks, standard safety precautions must be observed.

4.2 Catalysts

Production of flexible PUR foam requires the use of catalysts because two major reactions take place. In the polymerisation (or gelling reaction), polyfunctional isocyanates react with polyols to form PUR. In the gas-producing (or blowing) reaction, the isocyanate reacts with water to form polyuria and carbon dioxide. These reactions occur at different rates, with both reactions dependent upon temperature, catalyst level, catalyst type and various other factors. However, to produce high-quality foams, the rates of both reactions must be controlled and balanced.

If the gas-producing reaction (blowing) occurs faster than the polymerisation reaction (gelling), the gas generated by the reaction may expand before the polymer is strong enough to contain it, and internal splits and foam collapse can occur. In contrast, if the polymerisation occurs faster than the gas-producing reaction, the foam cells remain closed, causing the foam to shrink as it cools. If these two reactions are balanced appropriately, uniform open cells will dominate the foam structure. Open cells offer little resistance to diffusion and cell pressures quickly equilibrate without significant foam shrinkage.

Tin catalysts, especially stannous octoate, are usually chosen as one of the catalysts in the production of flexible PUR foam. Tin catalysts strongly catalyse the polymerisation reaction. Polyols and formulations that permit a range of tin levels to be used without causing processing problems are desirable. Insufficient catalyst will lead to foam splits or possibly collapse if the polymer fails to gel sufficiently. Excessive catalyst will cause closed cells and shrinkages. The larger the latitude of catalyst processing, the better the chance foam manufacturers have of successfully producing flexible slabstock foam with the desired physical properties and without processing problems.

In contrast to tin catalysts, tertiary amines catalyse the gas-producing reaction. The residual catalyst escapes from the finished foam after production or is incorporated into the polymer structure. The various tertiary amines in use differ greatly with regard to catalytic activity and efficiency, so overall reaction rates can be optimised by using mixed catalyst systems. Although amines and tin compounds catalyse different reactions within the foaming sequence, they do not act entirely independently. Typically, each catalyst influences both reactions in the foaming process, and the ability of foam manufacturers to maintain an appropriate balance between the two foaming reactions can be greatly influenced by the composition and selection of specific catalysts.

4.3 Blowing agents

Water and methylene chloride are the blowing agents commonly used by foam manufacturers in foam formulations. Others can also be used, but some are banned due to environmental issues. Methylene chloride remains the predominant auxiliary blowing agent. Water is present in every foam formulation but has a maximum threshold limit due to excess heat generation and fire hazards. Water reacts with isocyanate to form compounds that remain in the foam and also carbon dioxide, which acts as a blowing agent. Auxiliary carbon dioxide can also be added as a liquid to augment the blowing portion of the foaming reaction.

4.4 Surfactants

Flexible PUR foams are made with non-ionic, silicone-based surfactants. Different grades of silicone surfactants are available to the manufacturer to meet specific needs. Silicone surfactants carry out five main functions:
- Reduction of surface tension for improved chemical affinity with polyol.
- Provision of film resilience known as 'self-healing in the bubbles'. It is the film resilience that prevents collapse of the foam during foam rising, until chain extension and cross-linking reactions of the polymer have progressed sufficiently for the foam to be self-supporting.
- Silicone surfactants provide control of cell size through the promotion of homogenous fine cells. This phenomenon is known as 'emulsification'.
- Silicones provide bubble breakage at full rise. If this action does not occur, the foam is bound to shrink during cure. In this case, the resulting foam is said to be 'dead' or 'closed'.
- Silicone counteracts the deforming effect of any solids added to the reacting system.

Of all these factors, the most important is stabilisation of the cell walls. Surfactants prevent the coalescence of rapidly growing cells until they have attained sufficient strength through polymerisation to become self-supporting.

4.5 Methylene chloride

Methylene chloride is a liquid with a low boiling point. It is used in foam formulations to assist in attaining densities and softness not obtainable by the use of the conventional blowing agent: water. Methylene chloride absorbs heat from the exothermic reaction of the foam formulation, vaporising and providing additional gases useful in expanding the foam to a lower density. These liquids are also commonly used for cleaning and flushing.

Methylene chloride acts as an auxiliary blowing agent by complementing the blowing effects of carbon dioxide generated from the water/TDI reaction. It also acts as a heat sink, which means that a higher total catalyst package is needed to maintain adequate cure for the foam. These are liquids available in steel or plastic drums, and standard precautionary measures should be followed.

4.6 Additives

Additives are materials introduced into a foam formulation system to achieve specific or desired properties for particular end uses. They do not interfere with the basic foaming chemistry. Some of the most common additives are:
- Pigments
- Fillers
- Flame retardants
- Antioxidants
- Cell openers
- Plasticisers
- Anti-bacterial agents
- Anti-static agents
- UV stabilisers
- Colourants
- Foam hardeners
- Cross-linkers
- Compatibilisers

4.6.1 Pigments

Foams made from general foam formulation systems are colourless or 'white'. Pigments are added to foam systems for easy identification of products. This is a big help or maybe considered a must for large-volume foam manufacturers, where many types and grades of foam are made. Yellow is the preferred colour for hiding UV action on foam, and coloured foam also help the aesthetics of the product. Typical problems that maybe encountered by using pigments are:
- Foam instability
- Foam scorch
- Colour migration
- Limited colour range
- Abrasive action on pumps and mixers

Polyol- and water-based pigments are preferable. If water-based pigments are used, care must be taken to know the water content so as to make a suitable adjustment in the formulation to be under the permissible threshold for water content and thus prevent a fire hazard.

4.6.2 Fillers

Fillers are usually finely divided inert inorganic compounds and are added to foam formulations to increase density, load-bearing and sound absorption, and also result in cost reduction for the foam. The use of fillers may affect certain physical properties of foams.

Of all the wide range of fillers available, only the inorganic calcium carbonate fillers are most widely used in foam production. If fillers are used with diligence in foam formulations, substantial cost reductions can be achieved. However, care must be taken to use very dry fillers or at least to know of any moisture content factor to make the necessary adjustments in the formula so as not to exceed the allowable maximum water content.

4.6.3 Retardants

Low-density, open-celled flexible PUR foams have a large surface area and high permeability to air. Thus, they will burn given, a sufficient source of ignition and oxygen. Flame retardants are often added to reduce this flammability, as measured by small-scale laboratory tests. The choice of flame retardant for any specific foam is often dependent upon the intended service application of that foam and the attendant flammability testing results governing that application. Aspects of flammability that may be influenced by additives include initial ignitability, rate of burning and smoke evolution. Many grades of flame-retarding agents are available from special chemical suppliers and the most used ones are the chlorinated phosphate esters, chlorinated paraffins and melamine powders.

4.6.4 Antioxidants

To prevent the degradation of polymers, colour and viscosity as well as to protect the foam during its use, lifetime antioxidants are incorporated into formulations. Three types of antioxidants – primary, secondary and multifunctional – are available and can be used alone or in combination. A manufacturer will have a wide choice of different grades of antioxidants for use to suit different end applications for foams.

4.6.5 Anti-static agents

Some flexible foams are used in packaging, clothing and other applications, where they are desired to reduce the electrical resistance of the foam so that the build-up of static electrical charges is minimised. This has traditionally been accomplished through the addition of ionisable metal salts, carboxylic salts and phosphate esters. These agents function by being inherently conductive or by absorbing moisture from the air. The desired net result is achieved by the surface resistivity of the foam.

4.6.6 Cell openers

In some PUR foams, it is necessary to add cell-openers to prevent cells from shrinking upon cooling. Additives used for inducing cell opening are silicone-based additives, waxes, finely divided solids, paraffin oils, long-chain fatty acids and certain polyether polyols made using a high concentration of ethylene oxide.

4.6.7 Plasticisers

Non-reactive liquids are sometimes used to soften foams or to reduce the viscosity for ease of processing. The softening effect can be compensated for using a polyol of lower equivalent weight so that a higher cross-linked polymer structure is obtained. These materials may increase foam density and often adversely affect physical properties.

4.6.8 Anti-bacterial agents

Under certain conditions of warmth and humidity, PUR foams are susceptible to attack by microorganisms. If this is a concern or is requested by customers, anti-bacterial agents such as yeast or fungi are added during manufacture.

4.6.9 UV stabilisers

All PUR based on aromatic isocyanates turn dark shades of yellow upon aging with prolonged exposure to light. This yellowing is a surface phenomenon that is not a problem for most foam applications. Light protection agents such as zinc dibutyl thiocarbamate, hindered amines and phosphites can be used to improve the light stability of PUR, or colouring the foam yellow is also effective.

4.6.10 Colorants

Most flexible PUR foams are color-coded during manufacture to identify product grades, conceal yellowing and make an appealing consumer product. The usual practice is to add small amounts of pigments or dyes into a polyol mix. Typical inorganic colouring agents include titanium dioxide (white), iron oxide (red) and chromium oxides. Organic pigments originate from azo/diazo dyes, dioxazines as well as carbon black. Problems encountered with these colourants are high viscosity, abrasive tendencies, foam instability, foam scorch, migrating colour and a limited range of colours available. Most of these problems can be overcome by the use of correct amounts of chemicals, good dispersion in the polyol and process control.

4.6.11 Colour basics

The general colouring system consists of five basic colours:
– White
– Black
– Yellow
– Blue
– Red

Many other colours can be obtained by mixing two or more of the basic colours.

For example, black + white = grey
yellow + blue = green
red + white = pink
blue + red = brown

By mixing more than two, several other colours/shades can be obtained.

4.6.12 Foam hardeners

They are highly viscous polyol-based additives designed specifically for the production of high load-bearing flexible foams. They are polyether or polyester polyols with high functionality cross-linkers and copolymers to improve foam hardness and reduce the production cost while maintaining good physical properties. The use of foam hardeners can increase foam hardness by 50%, ensuring that less TDI is used and providing an additional blowing effect (thereby reducing the density).

4.6.13 Cross-linkers

Cross-linkers and chain extenders are, in general, low-molecular-weight polyols or polyamines used in a formulation to improve foam properties. They also help in the curing process.

4.6.14 Compatibilisers

Compatibilisers are molecules that allow two or more non-miscible ingredients to come together and give one clear homogenous liquid phase. Many such compounds are known to the PUR industry, such as amides, amines, hydrocarbon oils, polybutylene glycols and ureas.

4.7 Accessories

In the manufacture of flexible PUR foams, irrespective of the method used (manual, semi-automatic or continuous), there are certain important accessories needed for successful production.

4.7.1 Kraft paper

Kraft paper is used mainly for lining the bottom and sides of a conveyor belt for the continuous production of flexible foam slabstock. The paper should be of good quality, and coated or glazed paper is preferred to the non-coated for reasons of non-adherence and easy release. Good-quality paper has good dimensional stability, high stiffness and excellent flexibility when in use. Also, the paper must be of high density and low porosity to prevent penetration of liquid chemicals. These papers come in very large rolls and different widths to suit different machines and, being very heavy, need a suitable device (e.g. forklift) for transport.

4.7.2 Machine-glazed paper

A machine-glazed paper is a Kraft paper usually in the range 20–100 g. It is produced on a paper machine with a 'yankee cylinder' as the major drying unit. Hence, the paper is also stretched while drying, thereby giving stronger paper compared with a normal sack Kraft manufactured on a machine with a conventional drying unit. These papers have one side glossy and the other side is rough.

4.7.3 Polyethylene-coated paper

Polyethylene-coated Kraft paper is a machine-glazed or mono-finished paper with a non-peelable polyethylene film. For foaming applications, the weight is approximately 60–80 g with a polyethylene coating of 15–20 g. The advantages of using this paper are better overall strength (because it is a composite material), no soaking or seeping of rising foam, less frictional resistance to rising foam, formation of smooth and thin skin in contact areas (particularly the sides), easily removable due to non-adherence and lower weight per roll of paper.

4.7.4 Peelable Kraft paper

Peelable Kraft paper is a special grade of paper that is unglazed or unbleached and manufactured with a coat of low- or high-density polyethylene. This peelable film allows the foam producer to recover the paper only, leaving the film on the foam block surface. This quality of paper can be used on top and bottom as well as on the sides. This innovative product enables a foam producer to achieve smooth surfaces with less waste, resulting in important cost savings.

4.7.5 Plastic films

For a small-volume producer or a producer using the intermittent box method (where flexible foam blocks are made one by one), the use of low-cost plastic films to line the bottom and sides of a mould is sufficient and cost-effective. Usually, thin sheeting of polyethylene films is used to significant effect. However, care must be taken to eliminate crease on the film, which allows the rising foam to creep in and increase wastage.

4.7.6 Mould release agents

Except for the continuous foaming process, an alternative to using Kraft paper is mould release agents. Many grades are available, and the most commonly used ones are silicones and waxes. A foam manufacturer could also use a combination of polyethylene film and a suitable release agent.

4.8 Summary

Polyols are organic compounds with high molecular weight. In the manufacture of viscoelastic foams ('memory foam'), polyol blends are usually employed. A polyol blend consists of many materials. That is, a base polyol (making up the bulk of the blend), a chain extender polyol (a short-chain polyol that connects the longer chains of the base polyol) and a filler polyol (which makes the blend cheaper and also increases the foam density). Polyols have more than one hydroxyl group (OH).

Polyols are the major portion of a formulation, so selecting the correct polyol grade as the base is important. This is based on the final properties desired for the end applications. The most commonly used polyols for flexible PUR foams are the polyether polyols.

What differentiates polyether polyols from the others? Some of the more important parameters of a polyether polyol are:

– Molecular weight: sum of the atomic weights of the molecule (range, 150–12,000).
– Functionality: the number of reactive groups attached to polyol molecule is 2–6.
– Hydroxyl group content: the combined oxygen and hydrogen reactive radicals.
– Viscosity: thick/thin property of liquid affecting flow and processing.
– Compatibility with catalysts and other additives during foaming.

Isocyanates – the most commonly used ones are TDI and diphenylmethane diisocyanate. They are organic compounds of lower molecular weight than polyols and form exceptionally strong cross-linked structures with polyols (almost as a superstructure).

Water is generally used as a blowing agent to produce gas in flexible PUR foams.

Methylene chloride is also used as an additional blowing agent to increase gas content, thereby enabling foam manufacturers to produce very soft foams. Without blowing agents, foams would be hard.

Amine catalyst energises the water/TDI reaction to release carbon dioxide gas and helps the foam to rise.

Tin catalyst helps to energise the polyol/TDI reaction.

Surfactants are used to control the formation and structure of foam cells.

Air is used to enhance foam cell structures.

Colorants are pigments or dyes in powder or liquid form. They are yellow to counter degradation due to UV action. Other colours are also used in foams for easy identification of different densities, properties or customer preference.

Graft polyols are used in a 'polyol system' to reduce costs, increase hardness/density or texture when formulating, and for load-bearing applications.

Calcium carbonate is used as a fine powder to reduce foam costs. It may affect some foam properties.

These raw materials can be purchased separately, and foam manufacturers can carry out their own formulating to work out suitable systems for a particular customer or for the marketplace. This gives worthwhile cost savings but the manufacturer must have the skills to formulate correctly to achieve good foam structures and the desired properties for the end applications. This also gives the manufacturer the advantage of formulating for different densities, qualities and grades with the same raw materials. This may cause wastage, hazards and poor-quality blends due to possible mistakes in weighing and mixing the different components.

Alternatively, a foam manufacturer could request professional suppliers/blending companies dealing with urethane raw materials to supply blended systems of two to three components, which are easy to store and use but would cost more. In the case of conventional foams, these extra costs could be a factor but, in the case of viscoelastic or memory foam products, because the market prices for end products are high, it is very viable. In this case, due to high profit margins, pre-blended material systems would be better and safer methods of securing the raw materials. However, a disadvantage with blended and ready-to-use systems would be that only one density could be produced and the manufacturer would have to have several different systems in stock to meet the demands of different customers unless these can be purchased on a just-in-time basis. Here, it is important to note that the general shelf-life of most urethane chemicals is approximately 6 months.

Polyols and isocyanates are available in 45/55 gallon steel-coloured drums for easy identification and the other chemicals are available in smaller drums or packs for the small-to-medium manufacturers, but large-volume manufacturers may opt for drums, totes or bulk in tankers. The last option will ensure cost savings as well as the availability of raw materials for production at all times and avoid delays in supplies due to various reasons. However, the layout of such an operation incurs appreciable initial costs, and safety factors must also to be taken into account.

5 Properties and foaming technology of polyurethane foam

5.1 Properties of polyurethane foam

The properties of all water-blown conventional polyether polyurethane (PUR) foams are controlled by formulation and processing conditions, more so than individual component properties. Catalyst and surfactant effects are generally small for good-quality, high airflow cushioning grades used for furniture, bedding and carpet pads. Therefore, only the major effects of water, isocyanate (index) and polymer solids need to be considered. The term 'polymer solids', is used interchangeably with 'parts of polymer polyol', even though these two items differ by a constant factor. Processing conditions include those that affect foam breathability, structure and cell count. The critical properties of foams are density and indentation force deflection (IFD). The general rule of thumb is that, as water content increases, density and IFD decrease. Also, higher contents of isocyanate in a formula will give softer and lower densities and vice versa. This phenomenon is dependent on the amount of polymer polyol used and the actual amount of blowing agents (water +) in the formulation. At densities corresponding to 2.5–4.0 php water, IFD increases with increasing water. If more than 30 parts of high solid polymer polyols are used, foam firmness decreases with increasing water content. These effects are associated with water, polymer polyols and isocyanates.

Indices are results of linear regression analysis of data generated using designed experiments, which serve to separate interactions between direct variables. Density is modelled using its physical definition: density = mass/volume. Use of such a definition as a modelling base could allow more accurate extension to identify ranges beyond the actual data range. If very accurate calculations of IFD are desired, they can be obtained using linear statistical modelling methods. However, purely empirical models can be used confidently over only a specific data range from which the model was developed. Also, the concentrations of catalysts and surfactants are not considered in the work, so only quality 'open' foams can be predicted. Foams that are 'artificially' hard due to low breathability will not correlate to this modelling method.

5.1.1 Basic physical properties

Density – The density measurement is an apparent or bulk density, not the true polymer density. This physical property is important because of its parallel relationship with cost and load-bearing. In general, high densities result in higher costs and improved load-bearing properties. The density of a foam sample is calculated by careful and accurate measurements and weights. It is normally expressed as pounds

https://doi.org/10.1515/9783110643183-005

per cubic foot or kilogram per cubic metre and is the sample weight divided by its volume.

Compression – Compression force deflection (CFD) is similar to the IFD test in that the force required to compress the foam to a specified level is measured and recorded. The CFD test uses a plate large enough to apply uniform force across the entire top surface of a 50 × 50 × 20 mm foam sample. The foam sample is placed on a support plate, which has been perforated to allow airflow. The compression foot is brought into contact with the foam sample and the sample thickness measured at a contact force of 140 Pa. The sample is conditioned by compressing it twice to 25% of its original thickness. After the sample relaxes for 6 min, the force is applied to the compression foot so that the foam sample is at a constant rate. When the sample is compressed to 50% of its original height, this thickness is held for 60 s and the final load is recorded. The test is repeated several times and the average is taken as 50% CFD.

5.1.2 Tension test (tensile strength)

The tension test measures the strength of the foam under tension and gives information about elasticity. Samples for this test are die-cut and are shaped as a 'dogbone' or 'dumbell'. The sample is pulled at a constant rate until it breaks. This is done on a machine, and the force recorded at the breaking point is the tensile strength of the foam and is usually expressed in kPa. The maximum extension of the foam sample as a percentage of its original length is called the 'elongation at break'. Sometimes, errors may occur if actual contact is made by automatic machines due to sample deformation. However, if non-contact machines based on laser technology are used, these errors can be eliminated.

The modulus of elasticity is obtained by plotting the stress force against strain in a tension test. This modulus is the initial slope of the stress/strain straight line. Most PUR foams have a 'yield point', that is, the point at which deformation is no longer reversible and non-elastic deformation starts. This irreversibility may be due to the tearing of the struts in the foam, breaking of cell windows and the slipping of the polymer chain. The stress measured at the yield point is called the 'yield strength'.

5.1.3 Tear resistance

Another measure of the strength of foam is its tear resistance. In this test, a sample is slit at one end and the force required to tear it at a constant rate is measured. Results from this test are highly rate dependent. Tear strengths are not necessarily related to tensile strengths in foams. Results may differ with different shapes of foam samples.

5.1.4 Airflow

Airflow is a measurement of the openness or porosity of a foam. Airflow values are very sensitive to the orientation of the sample and two values are typically recorded: airflow parallel to foam rise and airflow perpendicular to foam rise. In moulded foams, airflow through the surface skin is also of interest.

5.1.5 Resilience

The resilience of a foam is measured in a ball rebound test. This consists of a dropping a steel ball on a foam sample and measuring and recording the height of the rebound. For this test, a simple testing machine or an automated one can be used.

5.1.6 Fogging

Fogging is an important criterion for foams used in automobile applications. Fogging refers to the problem of a light-scattering film that forms on interior glass surfaces. In a typical testing procedure, a foam sample is heated in an apparatus to a desired temperature to promote the release of volatile components. These components are then condensed on a cooled glass plate and subjected to a gravimetric or light reflectance loss of quantification. Generally, amines and plasticisers cause these problems, and suitable grades should be selected to give results within accepted tolerances.

5.1.7 Durability

Several methods can be selected for measuring fatigue in flexible foams. For the most part, they will fall into two categories: static fatigue tests and dynamic fatigue tests. Within a category, the individual tests vary in terms of degree of loading or deflection, use of indentation or full-size deflection, addition or absence of shear, sample thickness, number of cycles, rate of application of flex, and recovery or rest period after flexing. The sample type also varies. Measurements from the testing vary considerably. They include height loss, load loss, physical appearance or loss of other properties.

5.2 In industrial applications

Foam is widely used as an insulating material, especially in the construction industry. The R-value of an insulating material is a measure of its thermal resistance to heat. The higher the value, the greater is the resistance:

- The *R*-value of insulation materials is dependent upon ambient temperature and wind conditions. Independent tests show that at 8.5 °C with a 15-mph wind, the theoretical *R*-value of PUR foams drops from 19 to 18, while butt insulation drops from 19 to 7.
- In retrofits with smaller existing framing sizes, buildings can be insulated to meet the current code requirements.
- In new constructions, smaller framing sizes (lower lumber costs) can be insulated to current energy-efficiency standards.
- Plumbing can be installed on outside walls without freezing because only a thin layer of foam is required between pipes and external sheathing.

5.2.1 Foam is a good air sealant

Foams are effective air-sealant materials. That is, they provide an air barrier to prevent air movement through a building:
- Air leakage is the primary cause of poor building performance. Foam-insulated homes outperform conventionally insulated homes without requiring complicated and labour-intensive air-sealing work.
- Because foam is airtight, it performs better in windy conditions and resists loss in the *R*-value.
- Air leakage at penetrations creates an environment for condensation. This affects the overall performance and can compromise indoor air quality (bugs, mould and rot). Condensation can also lead to premature structural failure in structural framing and sheathing materials.
- Independent testing has shown that PUR foam-insulated buildings can perform well above current energy standards.

5.3 Closed-cell foams

Foams with closed cells have very low air- or water-vapour-transmitting properties and are widely used in construction of buildings as barrier materials:
- Closed-cell foams have a very low permeance (potential for water vapour to pass through them).
- This provides protection against moisture transport into the insulation with its related potential for condensation. Vapour that remains on the inside (warm side) will not come in contact with cold surfaces, where the dew point can be reached.
- Imperfections in vapour retarders are less critical with closed-cell foams.
- Indoor humidity levels are more easily maintained at healthy levels if vapour cannot escape during winter weather.

- PUR foams are not damaged by roof leaks, foundation leaks or condensation.
- PUR foams can be used in masonry construction and not sustain damage from water penetration.
- PUR foam sealants can help protect structures against wind-driven rain penetration.

5.3.1 Foams have structural advantages

All foams are used widely in the building construction industry due to their distinct property advantages with regard to structural work, ease of installation and low weight:
- Foams can help to resist wind shear.
- Foams can reinforce exterior sheathing and windows.
- PUR foams are used in structural panels and other composite structures.
- Foams can be walked on or nailed into without damaging performance. They can also be washed without damage.

5.3.2 Foam usage in sound control

The versatile properties of foams makes them ideal for sound-proofing, sound-damping and suppression of air-borne noise, especially in industrial applications, theatres, offices, sound studios and many other applications:
- Open- and closed-cell foams provide good sealing against airborne sound transmissions.
- Open- and closed-cell foams provide good sound-damping against airborne sound transmission.
- No low-density insulation materials are effective against structure-borne sound. Double-layer structural systems and resilient structural materials are probably the best defence against structure-borne sound.

5.4 Foaming technology

Flexible PUR foam manufacturing technology is quite complex. Thorough knowledge and experience is needed to formulate correctly to obtain quality production. The primary source of know-how is with the manufacturer or a member of his/her staff who has been trained or has previous experience. The secondary source is the supplier of the raw material, who can provide a wealth of information and datasheets on all products purchased. Nevertheless, it is the responsibility of the producer to sift through all the data available and to formulate correctly to meet the various customer demands.

The availability of a small or well-equipped in-house laboratory will help tremendously. This will enable a foam manufacturer to try out small batches in the laboratory and make the necessary adjustments and improvements before attempting large-scale production runs which could result in waste. Formulating for viscoelastic ('memory' foam; see Chapter 1) with the most important factors (density, support factor or whether 'slow' or 'fast' recovery is desired) could pose problems initially. Hence, small-scale laboratory tests will help. In reality, the more important features for a manufacturer/formulator to thoroughly understand are the functions of each raw material, rather than the complex chemical formula for each raw material. There are plenty of organisations where theoretical and practical training can be used in the fields of raw materials, processing machinery and testing methods. For a small-to-medium producer, it is best to outsource the high-technology testing and certifying of products unless they can do so themselves in-house.

5.4.1 Raw materials and their functions

The essential components for the production of flexible PUR foams are:
-   Polyols
-   Diisocyanate
-   Water
-   Auxiliary blowing agents
-   Catalysts (tertiary amines and tin salts)
-   Silicone surfactants
-   Graft polyols
-   Extenders
-   Colourants
-   Fillers
-   Additives

5.4.2 Polyols

Most flexible PUR foams are made from polyether polyols. Essentially, they are propylene oxide and ethylene oxide copolymers with a trifunctional initiator, and therefore are triols. The polyols for standard block foams have hydroxyl values in the range 46–56 mg KOH/g (nominal molecular weights of 3,500–3,000, respectively). Some formulations have more than one polyol and form a blend. In a formula, the total content of polyol will be the largest portion, and will be taken as 100 parts by weight. All other components are calculated in relation to this base number.

5.4.3 Diisocyanate

The most commonly used diisocyanate for flexible foams is toluene diisocyanate (TDI). Commercial grades of TDI are mixtures of the 2,4 and 2,6 isomers in controlled proportions. The most frequently used grade to manufacture flexible PUR foams consists of 80 parts of the 2,4 isomer and 20 parts of the 2,6 isomer. Diphenylmethane diisocyanate is also used for making flexible foams, especially in moulded flexible foams. The TDI component is the second highest component in a formula.

5.4.4 Water

Water is the primary blowing agent and reacts with the diisocyanate to generate carbon dioxide gas, which causes the foam to expand. Without this gas the foam would be hard. There is a maximum limit of 5 pbw in relation to the 100 pbw of polyol but some formulations may have slightly higher contents. This threshold prevents possible fire hazards.

5.4.5 Auxiliary blowing agents

An auxiliary blowing agent (e.g. methylene chloride) is used in combination with water to produce foam densities <21 kg/m^3 or to produce soft foams at all densities. These auxiliary blowing agents are liquids with low boiling points.

5.4.6 Catalysts

Tertiary amines energise, accelerate and control the rate of the water/diisocyanate reaction. Tin salts are specific for the reaction between polyol and diisocyanate. The almost universally used catalyst is stannous octoate (often referred to as 'tin catalyst'). To control the small quantities required, stannous octoate is diluted with polyol before use or can be purchased in an already diluted form. Some manufacturers may prefer to purchase this component in pure form. An alternate catalyst for use, especially if polyol pre-blends (containing water, amines and silicone) are made, is dibutyl tin dilaurate.

5.4.7 Silicone surfactants

A surfactant is essential to control the foaming process. In the manufacture of flexible PUR foam, the most commonly used one is a silicone surfactant. Its two main functions are (a) to assist the mixing of the components to form a homogenous liquid and (b) to stabilise the bubbles/cell structure in the foam during the expansion to prevent collapse before the liquid phase has polymerised.

5.4.8 Graft polyols

Graft polyether polyols contain copolymerised styrene and acrylonitrile with low-inhibition compounds designed to maximise load-bearing properties in foam applications such as carpet underlay, furniture foam, bedding and protective foams. These special polymers used to have a solids content of 10–15% but now some well-known manufacturers are offering graft polyols with 45% solid content. These polymers will produce cost-effective foams.

5.4.9 Extenders

An extender is a special type of polyol of many short chains that connects the longer chains of the base polyol or blends. Chain extenders/cross-linkers are usually low-molecular-weight polyols or polyamines used to improve the properties of PUR foam by aiding the curing process.

5.4.10 Colourants

In large-volume production plants where many different types of foams (e.g. those with different densities and different physical properties) are produced, pigments or dyes that are compatible with polyols are used to colour the foam for easy identification, aesthetic features or as the customer requests. Smaller-volume foam producers where only one or two qualities of foam are produced may opt to colour their foam blocks yellow to counter the action of ultraviolet (UV) light.

5.4.11 Fillers

Fillers are incorporated in a formula to reduce costs. Of the many fillers that can be used, the most popular and widely used is calcium carbonate in fine powder form. Excessive use could affect the physical properties of the foam.

5.4.12 Additives

The incorporation of additives is to protect the foam and improve the foam quality. The two most incorporated additives in the manufacture of flexible PUR foam are flame retardants and anti-UV agents. Other additives are to counter oxidation and microorganisms.

5.5 Foaming process

The sequence of events described below is a simplified picture of the manufacturing process of flexible PUR foam and is basically a mixing and dispensing operation.

5.5.1 Mixing

In the first step, the formulated ingredients are mixed in sequence by means of an efficient mixing head. Good mixing is essential to produce a homogenous liquid mix, which in turn produces good-quality foam with fine cell structures. The silicone surfactant assists in achieving good mixing because it also lowers the surface tension of the polyol.

5.5.2 Nucleation

During mixing, air bubbles are created in the liquid mix. These act as a nucleation point for the expanding gases. When making foam using the box method and with simple equipment, it is not always possible to control the number of bubbles or the desired cell size. However, in continuous slabstock machines, there are several ways to ensure that there are sufficient initiating points of foam formation to achieve controlled and uniform cell sizes as desired. After approximately 10 s, the blowing gases – carbon dioxide and an auxiliary blowing agent (if used) – diffuse into small air bubbles and then enlarge them, giving the liquid mix a 'creamy' appearance. The time from initial mixing to this change in appearance is called the 'cream time'.

5.5.3 Expansion

As more gas is generated, the bubbles expand and the foam begins to rise. During the rising of the foam, the number of bubbles remains constant. The silicone surfactant stabilises the bubbles, preventing them from coalescing; without the surfactant the mixture would boil and the foam collapses. At the same time as the bubbles are expanding, a polymerisation reaction takes place after mixing and the gas reaction stops. At this stage, the foam mass will occupy about 30–50 times the original liquid volume. The polymer part of the foam has begun to gel in the form of gas-filled cells with thin walls and thicker 'struts' along the edges.

In foam formation, the two types of reactions – one causing gas generation and the other, the polymerisation of the polymer cell walls and struts – must be finely balanced. If the balance is incorrect, various faults can occur in the foam. The two main types of faults are caused by the polymerisation reaction occurring too early or too late. If it occurs too soon, polymer strength develops too early and some of the cell walls will not burst under the gas pressure. Hence, the foam will have poor resilience and feel 'dead' (and is often called 'tight' foam). If many of the cells remain closed, then if the foam cools and the internal gas pressure falls below the atmospheric pressure, the foam will shrink.

If the polymerisation occurs too late, the struts will be weak at the time when the cell walls burst. If this happens, the struts can break and because the struts of a cell are shared by all surrounding cells, the result is a series of struts breaking, thereby forming a split. The extent of this split can be increased by the expanding gases, forcing the split further apart and breaking even more struts. Splits due to polymerisation problems usually occur at the top edges (shoulders) of a block where the foam is weakest (coldest and therefore slower to polymerise) and in the centre of a block, where the exothermic is highest (highest gas pressure).

5.5.4 Cell opening

In flexible foams, the cell walls cannot contain the full gas pressure, and at the same full rise time, they break. The polymer is gelled sufficiently for the struts to be strong enough to stand while the gas escapes through the open cells.

5.5.5 Gelation

Continuing polymerisation increases the strength of the polymer, which reaches the full gel time a few minutes after mixing.

5.5.6 Curing

In flexible foams, curing takes places in two stages. The first curing is when the gas reactions have stopped and an indication will be when the top skin of the foam becomes tack free.

Depending upon the formula and ingredients used, this period can vary slightly and is approximately 8–10 min. In continuous foaming, this time could be even shorter. The foam in the form of blocks, buns or round blocks is transported to a holding area away from the production floor and left for 24 h for the second curing stage, during which time various slow cross-linking reactions take place with the emission of heat due to the exothermic reactions to give the foam its final physical strength.

5.6 Storage of foam blocks

For fire-hazard and safety reasons, the storage area of semi-cured foam blocks should be in a holding area far away from the storage area of raw materials as well as the production area. This area should be well ventilated and with an entrance and an exit with easy access to the next stage of the process, that is, cutting/ fabrication. It is best to have a storage system where a first-in/first-out system is implemented to ensure the correct blocks are taken for the next stage after a 24 h final cure period. Exothermic reactions are taking place in these blocks, so care must be taken to store them ≥1-foot apart and without stacking them one on top of each other to prevent possible fire hazards. In the case of viscoelastic blocks, the final curing time could be ≥48 h. Due to the very heavy nature of these blocks, special transportation devices may be needed.

5.7 Cutting and fabrication

The foam blocks now undergo a sequence of operations before the final products are realised. First, the thick/rough outer skins (top, bottom and sides) are removed using horizontal and vertical band-saw cutting machines that may have different types of blades. These machines (other than the manually operated ones) are easy to use because they have semi-automatic and automatic electronic controls and provide great accuracy. Once the skins have been removed, a smooth foam block is available for cutting or fabrication of the end products desired. Many types of machinery (e.g. peeling machines, convoluting machines, laminators, contour cutters, edge-rounding and slab cutters) are available for a foam manufacturer to produce any desired end product. The wastes from production and block skins may be sold for packing, but the major portion of foam waste goes into shredders and size-reduction machines. They are rebounded into sizable blocks and cut again into thin sheets for carpet underlay or thicker sheets for the bases of foam mattresses. Saleable foam sponges of different shapes and sizes can be made from large pieces of foam scrap.

5.8 Filled foam

The demand for filled foam has been increasing in some parts of the world. It has been produced quite successfully on conventional machines but with limitations of height. The use of inorganic fillers is an important method of producing cheap-quality foam without too much reduction in foam properties provided it is well formulated and processed appropriately. Years ago, producers were satisfied with block heights of 26 inches or 66 cm but with new technology (e.g. Maxfoam machines), a foam of 42 inches ($\approx$107 cm) with 100% filled polyol. In general, the production of this type of foam is not difficult provided certain basic conditions are followed.

5.8.1 Type of filler

Fillers must meet the following requirements:
- The particle size must be $\leq$5 µm.
- The water content should be $\leq$0.2%.
- The filler used should not be hygroscopic.
- The filler should not contain free metal parts (which destroy catalyst activity).
- The filler must be easily dispersed in the polyol. In a weight ratio of 1:1, the slurry must have a low enough viscosity to be pumped. One of the inorganic fillers that meets all these requirements is barium sulfate. Producers may opt to use barium sulfate instead of calcium carbonate due to a higher specific gravity of 55%.

5.8.2 How much filler can be used?

In general, up to 100% filler can be used. More important is the water content/ auxiliary blowing agent level in combination with the filler percentage/block height because these four factors determine the limit in which the production becomes critical. In the machine, the fallplate setting determines whether fault-free foam can be produced.

5.9 Effects of changes in physical properties

If conventional polyether flexible foam is compared with this foam, the effects of changes in physical property are:
- Elongation at tear strength reduced by approximately 15–20%.
- Tear strength reduced by almost equal to 5%.
- Increase in compression set by approximately 300%.

5.10 Processing difficulties

Production of foams is a combination of several chemical reactions. Certain features highlight the problems that could occur due to faulty formulating, mixing or malfunctioning equipment:
- Very narrow spread of tin due to increasing filler content.
- Mechanical faults can be made very easily in this type of foam due to different rise patterns. The catalyst level on a highly filled foam is higher than on a conventional foam, and a much longer rise time should be expected. At the time of rise to full height, the foam is very delicate and can be damaged very easily. Therefore, a good fallplate setting which follows downward expansion of the foam and a smooth lay down of the foam mix is essential.
- The TDI index should be >106.
- A homogenous slurry, together with a narrow range in specific gravity, is desired.
- The water content of the slurry should be established before production.
- Preferably, a 'box' test should be carried out in the laboratory before production.

5.11 Description of additional equipment

To produce filled foam, the following additional items are required:
- A batch or slurry tank: Depending on the duration of the run, the tank volume must be determined. It is preferred to use a closed tank with a removable lid on the top. The filler must be emptied very carefully through a sieve into a predetermined quantity of polyol and, during the preparation, the slurry must be circulated through the metering circuit and the agitator must be working at a maximum speed of approximately 50 revolutions per minute (rpm). As soon as all the filler is mixed in, the agitator speed can be reduced to 15–20 rpm with the metering pump running and the tank lid closed.
- Recommended agitator specifications are:
  - Two-speed motor 8–10 HP
  - Low speed 15–20 rpm
  - High speed 50–60 rpm
- The propeller can be made of metal sheets that are firmly connected to the axle with a clearance to the tank walls of ≤30 cm. The lower propeller should be close to the bottom and the second one is approximately 90 cm from the bottom. The outlet opening of the tank into the feed line of the pump must be provided with a coarse grid or sieve. The return line must be at the top of the tank to prevent splashing.
- Metering pump for polyol/filler mix: Accurate metering is required at an output rate of almost equal to 250 kg/min. To extend the life of the pump, small parts of non-hydrolysable silicone oil may be mixed into the slurry.

- Metering lines and filters: A 4-inch suction line and a 3-inch delivery line with an oversized 6-inch filter in the pressure line can act as a filter.
- Temperature control: The recommended temperature is approximately 35 °C. Higher temperatures cut back on the catalyst level but the major disadvantage of using heat as a catalyst lies in the fact that a long curing time results in a thick top skin and possible damage to the side skins. The friction heat developed during agitation for mixing raises the temperature of the slurry and can be checked with a pyrometer or probe. Automatic temperature control on the outside of the tank walls can also be used. Cooling on the outside and heat development on the inside will result in condensation if the slurry tank is open. Therefore, it is better to have a closed tank with a dry air supply on the top.

5.12 Procedure

The best way to establish a workable formulation is to conduct small-scale trials in the laboratory. Mixing the polyol and the desired quantity of filler 2 h before the trials will help to remove air trapped in the slurry. Catalyst levels can be as used in virgin foam formulations and adjusted later. If the density is within specifications and the IFD value is high, a small machine run is worthwhile and the optimum fallplate setting can be determined.

6 Foaming calculations, other calculations and formulations

A manufacturer must have a thorough understanding of how to calculate and formulate to make quality foams. In previous chapters, the raw materials have been discussed in detail. Once the functions of each component are understood, the next step is to formulate correctly to achieve good-quality foams with the desired mechanical and physical properties. The information provided here has been made as simple as possible but is more than adequate for successful formulating and, because it is a chemical processing, one may use it as guidelines and 'fine tune' if necessary.

6.1 Calculations

6.1.1 Density

In foam calculations, there are different parameters that are required for successful formulating. The density of the final product is one of the key factors:

Mass (weight) = volume × density

where mass is expressed in g or kg; density in mass/volume; volume in cm^3 or m^3. Thus, density is expressed in g/cm^3 or kg/m^3.

6.1.2 Mixing ratio

A mixing ratio for a two-component system of raw materials calculates the specific amounts of isocyanate and polyol blend needed for the reaction. To calculate the mixing ratio, one needs to know the following information given in the certificate of analysis sent by the supplier:

- The hydroxyl number or hydroxyl number of the polyol (component B)
- The amount of water as a percentage present in the polyol
- The isocyanate content [also called the nitrogen–carbon–oxygen group (NCO) content] of diphenylmethane diisocyanate (MDI) (component A)

$$\text{Amount of MDI} = \frac{\{(\text{polyol OH} \times 100) + (\% \text{ water} \times 6,233)\}}{\{\text{NCO content} \times 13.35\}}$$

The result is the amount of MDI to add to every 100 parts of polyol.

https://doi.org/10.1515/9783110643183-006

Example: A polyol blend has an OH number of 95 and water content of 0.45%. The MDI has an NCO content of 23%. What is the mixing ratio?

$$\text{Amount of MDI} = \{(95 \times 100) + (0.45 \times 6,233)\}/\{23 \times 13.35\}$$
$$= 12,305/307 = 40.1$$

Therefore, 40.1 parts of MDI are needed for every 100 parts of polyol.

6.1.3 Isocyanate index

The isocyanate index is very important for fine-tuning the mix. The mixing ratio gives the actual proportions of the two components for mixing. Some two-component systems give the best results using slightly extra MDI, whereas others give best results using a little less MDI. In these cases, a manufacturer will want to adjust the mixing ratio slightly. The isocyanate index is the number that gives the exact amount by which to adjust. When the two components are supplied, the supplier gives an indication of this index. For example, if an index number is given as 98, then it is recommended that the manufacturer should use 98% of the MDI as calculated. The mixing ratio would then be 98% of 40.1 or 39.3 parts of MDI for every 100 parts of polyol. If the recommended index is 103, a manufacturer should use 103% of 40.1 or 41.3 parts of MDI per 100 parts of polyol. Sometimes the index may be given as 0.98 or 1.03, which means the same thing, only that they have already been divided by 100.

6.1.4 Ratio calculation

In foam formulations, the two main components are the polyols and isocyanates. These often determine the final density of the foam, and it is important to calculate the correct ratio between them.

Example: Formula ratio of 100 parts of component A to 160 parts of component B (by weight).

Also, taking the actually metered weight of component A as 1,200 g/min, then to determine the throughput of component B to correspond to given ratio, then

(g/min Part A ÷ g/min Part B) = (100 ÷ 160) = (1,200 ÷ Part B)
Therefore, 100 B = 1,200 × 160
B = 1,920 g/min = throughput for component B.

6.1.5 Pump rates per minute versus flow rate

Other than the manual operation, all foaming machines have different chemical components in containers or tanks. These are connected to the mixing head via pumps and flowmeters for delivery to the mixing head. It is important to calibrate, set and carry out frequent checks to ensure the correct volumes of chemical flow.

Example: 900 g/min of component B was metered at 110 rpm.
    To determine the rpm of pump for the desired flow rate of 1,920 g/min of B:

Using a simple ratio: $\dfrac{\text{known rpm}}{\text{known flow rate}} = \dfrac{\text{rpm desired}}{\text{flow rate desired}}$

$$\frac{110}{900} = \frac{\text{rpm desired}}{1,920}$$

Therefore, 900 × rpm desired = 110 × 1,920
                                = 235 rpm pump speed for B

Note: These calculations are only a guide because pump flow rates are not always straight-line functions. Actual metering of component B is recommended.

6.1.6 Machine flow rate versus mould volume

Assumption:
–   Mould volume = 3 ft^3
–   Component ratio A:B = 100:160
–   By trials, this ratio has produced an end product having a density of 2 lb/ft^3
–   Cream time = 30 s
–   Mould should be filled in 15 s

Exercise: To determine the
–   Weight of chemicals required
–   Output of each component

Weight of chemicals = density of product × volume of product
                    = 2 lb/ft^3 × 3 ft^3 = 6 lb

Required machine output in lb/m = 60 s × 6 lb = 24 lb/min

Component A, lb per min:

$$\frac{\text{Part A}}{\text{Part A}+\text{Part B}} = \frac{\text{throughput of A in lb/min}}{24}$$
$$\frac{100}{100+160} = \frac{\text{throughput of A}}{24}$$

Therefore, throughput of A = (100 × 24) ÷ 260 = 9.23 lb/min.

Component B in lb per min   = (24 − 9.23) = 14.77 lb/min.

6.2 Calculations for making large foam blocks: discontinuous process

Standard mattress sizes can vary from single size to king size. The three main items for producing good-quality foam blocks are correct mould size, correct formulating and adequate volume of chemicals in liquid form to be poured into the mould. All these need careful and effective calculations.

Exercise: To make a large viscoelastic ('memory') foam block to cut standard-size mattresses.

Requirements:
–   Size 80 × 60 × 28 inches (height)
–   Density = 4.0 lb/ft^3
–   Raw material system = two-component system

The required mould sizes should be designed by including a shrinkage factor of 1/2–3/4 inch for viscoelastic foams on each side but are also dependent upon the formulation. This is when a laboratory would help to carry out a small 'box test' before production. The machinery suppliers provide one or two moulds with the foaming equipment, but a foam manufacturer would have to give specifications of the sizes required. Size-adjustable moulds would be a solution and, for a medium-volume producer, it would enable him/her to make different sizes of foam blocks with the same mould.

Hence, the minimum size of mould required to make a final block of size 80 × 60 × 28 inches would be 81.5 × 61.5 × 28 inches (height). Here, even a slightly higher (>28 inches) value for the height may be considered if a flat top cannot be achieved.

6.2.1 Calculation of material required

Calculation of the material required is important because a manufacturer must ensure that sufficient material is to be poured for the foam to rise, fill the full mould and also rise high enough to reach the desired height. If there is too much material, the foam will overflow over the mould and be wasted. When calculating the exact quantity of material required, a gas evaporation factor of around 0.5–1.0% (minimum) should be added. A small wastage factor may also be considered. Again, a small box test in the laboratory would be of great help. Selection of the dispensing machine will be greatly dependent upon the minimum shot size required; otherwise, the desired height cannot be achieved.

Then, using formula $M = V \times D$

where M shows the weight in lb or kg, V is the volume in ft^3 or m^3 and D is the density in lb/ft^{3}or kg/m^3:

$$M \text{ (weight)} = \{(81.5 \times 61.5 \times 28) \text{ inch}^3 \times 4.0 \text{ lb/ft}^3 \} \div 1,728 \text{ inch}^3$$
$$= 324.86 \text{ lb} = 147.7 \text{ kg} + 1\% = 149.2 \text{ kg}$$

Therefore, the dispensing machine must have a single shot capacity of a minimum of approximately 150 kg or the foam producer will have the option of making blocks of reduced heights because the other dimensions cannot be less. For viscoelastic foam blocks, due to the very heavy weight of a block, the recommended height is 24 inches but 28 inches is also possible.

Now, consider a two-component raw material system for making memory foam with the following data:
- Mixing ratio = component B (polyol) : component A (iso) = 100:31
- Cream time = 11 s
- Rise time = 143 s

Then
- Component B = 100/131 × 149.2 = 113.89 kg
- Component A = 31/131 × 149.2 = 35.31 kg

Therefore, a producer would mix 113.89 kg of polyol blend with 35.31 kg of isocyanate with expected times of a 11 scream time and a total rise time of 143 s. The density would be approximately 4.3 lb/ft^3 or 69.5 kg/m^3.

Exercise: To make a large flexible foam block to cut mattresses.

Requirements:
- Normal flexible foam suitable for mattresses.
- Block size 80 × 60 × 42 inches
- Density = ≈1.8 lb/ft^3 or 28.8 kg/m^3
- Standard raw material formulation
- Shrinkage factor = 0.5 inches on each side

Then, using formula M (weight) = volume × density

M = 81 × 61 × 42 inches/1,728 inch^3 = 120 ft^3

= 120 ft^3 × 1.8 lb/ft^3 = 216 lb = 98.2 kg

6.2.2 Calculating raw material components

Correct formulating is essential for the production of quality foam blocks, and many formulations are available from different sources. Table 6.1 shows the recommendations for a practical and workable formulation.

Table 6.1: Raw material component calculations.

Component	Part by weight	Proportion	Weight (kg)
Polyol	100.00	100/175.95 × 98.2	55.81
TDI	70.00	70/175.95 × 98.2	39.07
Water	4.50	4.5/175.95 × 98.2	2.51
Surfactant	0.60	0.6/175.95 × 98.2	0.33
Amine catalyst	0.30	0.3/175.95 × 98.2	0.17
Filler	0.00	0.00	0.00
Tin catalyst	0.20	0.2/175.95 × 98.2	0.11
Methylene chloride	0.35	0.35/175.95 × 98.2	0.20
Total	175.95	Total	98.20

Note: For greater accuracy, if needed, calculate the smaller quantities to the third decimal place and weight in grams. Hence, a foam producer will have a workable formula for this particular production. If a floating lid is not used to flatten the top, the foam height of the block will have to be increased to counter the rounded shape on top due to a meniscus forming during foaming.

6.3 Typical formulations

The following are fairly accurate formulations and familiar to practice by the author. However, they should be used only as a guideline because the processing conditions such as atmospheric/room temperature, processing equipment and processing methods will have an impact and some adjustments may have to be made to achieve the desired qualities:

- Conventional foam
- High-resilience foam
- Viscoelastic foam
- Slabstock foam

6.3.1 Conventional foam (density range 16.0–32 kg/m³)

These formulae can be considered for production of large flexible foam blocks for furniture and bedding. They suit the discontinuous method of making blocks one at a time either by the simple fabricated wooden box mould or by the more sophisticated batch foaming dispensing machines. Table 6.2 shows some recommended practical formulations.

Table 6.2: Typical formulations.

Component	Standard	Mattress	Pillows
Polyol	85.00 pbw	80.00 pbw	100.00 pbw
Graft polyol	15.00	20.00	–
Toluene diisocyanate (TDI)	40.70	70.00	84.00
Water	2.90	3.50	4.50
Surfactant	1.20	0.60	0.30
Amine catalyst	0.12	0.70	0.27
Tin catalyst	0.04	–	–
Methylene chloride	–	–	–
Flame retardant	3.00	–	–
	Foam properties		
Density (kg/m³)	27.2	20.8	17.6
Support factor (IFD)	2.3	2.1	2.0

Note: As a general rule, increases in water and TDI contents yields softer foams and vice versa.

6.3.2 High-resilience foams

Consider this water-blown high-resilience flexible polyurethane (PUR) foam system as an example. This system, like many others, can be manually drill-mixed (meaning using a suitable electric drill or similar device) or for medium-volume productions a multi-component PUR dispensing machine with a suitable shot capacity can be employed. This type of system would ideally suit an entrepreneur with limited knowledge/experience and a small budget. These data (Table 6.3) are typical for two-component systems but may vary with different suppliers of these systems.

Data: A polyol blend and an isocyanate as separate ingredients, comprising a system.

6.3.3 Viscoelastic foam

The general density range for these speciality foams is 48–96 kg/m³. A foam manufacturer would need experience and know-how to formulate by himself or herself.

Table 6.3: Reaction profile.

Mixing speed	3,000 rpm
Mixing ratio	Polyol:iso = 100:57
Mixing time (s)	7–8
Cream time (s)	12
Gel time (s)	70
Tack-free time	300
Free rise density (kg/m^3)	57

Note: By increasing or lowering the isocyanate content, a foam producer can make harder or softer foams, respectively. If colour is used, add a desired quantity to the polyol blend and mix well before incorporation of the isocyanate.

This is especially so, because unlike standard conventional foams, these foams have four-dimensional properties, which makes them much more difficult to formulate to achieve the desired final properties. However, two-component systems that are easy to process are available from well-known suppliers. Using two-component systems have a disadvantage in that one system can usually produce only one density. Hence, a foam producer will have to have different systems if different densities are required. For entrepreneurs and small-volume producers, it is advisable to use two-component systems even if it is more expensive because the returns are much more than for standard foams.

Consider the following two-component system that can be processed on high-pressure and low-pressure dispensing machines. If a laboratory test is desired before production, the two components can be mixed in the ratio of polyol blend:iso = 100:48 using a disc-type laboratory stirrer with a disc diameter of 60 mm at a speed of approximately 1,300 rpm for 7–8 s. The reaction characteristics of the mixture are influenced by the intensity of the mixing, so the specified values shown in Table 6.4 should be considered for guidance only.

Table 6.4: Mixing ratio for polyol blend:MDI = 100:48.

Mixing speed	1,300 rpm minimum
Mixing time (s)	7–8
Cream time (s)	10
Gel time (s)	110
Rise time (s)	170
Free rise density (kg/m^3)	53

Note: Unlike conventional foams in which 24 h is sufficient for the final cure, viscoelastic foam blocks after de-moulding may require more time for the final cure and could be ≤48 h.

6.3.4 Slabstock foam (continuous foaming)

There are many types of continuous foaming machines but the general processing principle is the same. Machine settings may vary according to the controls and the material feed systems. The following formulation and settings (Tables 6.5 and 6.6) are for a standard continuous foaming machine and may be used as a practical guideline.

Parameters:
- Foam product: flexible PUR foam (yellow)
- Production method: continuous foaming
- Density: 20 kg/m^3
- Foaming time: 10.15 am
- Foaming temperature: 22 °C
- Foam block: 1.08 m (height) × 1.92 m (width)

Table 6.5: Formulation of flexible polyurethane foam.

Component	pbw	Weight (kg)
Polyol	72.00	90.83
Graft polyol	28.00	35.32
Poly/calcium blend	10.80	13.62
TDI	59.84	77.56
Methyl chloride	6.00	7.80
Water	4.60	5.96
Silicone	1.30	1.68
Amine catalyst	0.23	0.30
Tin catalyst	0.28	0.37
Colour	0.10	0.13

Table 6.6: Machine settings: small trough-coated Kraft paper.

Air	2,500
Conveyor	4.21
Fall plate	729 mm
Fall plate	617 mm
Fall plate	458 mm
Fall plate	275 mm
Fall plate	50 mm

6.4 Calculating indentation force deflection

The indentation force deflection (IFD) value as a number is, in most cases, required as a marketing tool. This is the ratio of compression of a selected foam sample at 25% and 65%. The test is the compression of a foam sample of thickness 60 × 60 × 10 cm being compressed to 25% and 65% of the foam sample height using an indenter plate of an electromechanical device and noting the force required to do this. The ratio of these values (65%:25%) is known as the IFD value and a value of ≥2 is considered acceptable and <2 considered to be below standard.

The values of 25% IFD and 65% IFD are the forces required during the load cycle for deflection of 25% and 65%, respectively, of the original foam height. IFD return to 25% is the force measured during the unload cycle at the 25% deflection point. Percent hysteresis return (also called 'recovery') is IFD return to 25% and expressed as a percent of 25% as follows:

% Hysteresis = {(Return 25% IFD) ÷ (Original 25% IFD)} × 100

Two other factors related to IFD are the guide factor and support factor. The guide factor is the ratio of IFD to density. Foams with higher guide factors have cost advantages but not necessarily performance advantages.

25% IFD guide factor = (25% IFD) ÷ (density)
or
65% IFD guide factor = (65% IFD) ÷ (density)

The support factor is the ratio of 65% IFD to 25% IFD. This number gives an indication of cushioning quality. Higher support factors indicate better cushioning quality. This quantity is also known as the sag factor or as the modulus:

support factor = (65% IFD) ÷ (25% IFD)

6.5 Calculating electrical power

Electrical power is one of the key (if not the most important) factors in any manufacturing activity, especially if planning to set up such a unit. The following are guidelines for calculating electrical power.

Whatever machinery and equipment is used, the manufacturers in their specification sheets indicate the horse power (HP) of each motor or device and the voltage (e.g. a motor will be specified as a single-phase 1-HP motor or as a three-phase 5-HP motor). In a manufacturing operation, the main items that will need electrical power are motors, equipment and lighting. Power is generally expressed as kilowatts (kW) and electrical consumption costs are based on a basic unit – kWh – the quantity of electrical power consumed in 1 h. Voltages may range from 110/120 to 230/240 V depending on the location. When installing or requesting a power supply, an additional 40–50% should be built in to accommodate the start-up for motors. It is always prudent to give consideration for possible expansion programmes.

Amperes (I), when horse power (HP) is known for:
- Single phase = (HP × 746)/(E × Eff × pf)
- Three phase = (HP × 746)/(1.73 × E × Eff × pf)

where I represents the current in amperes, E the voltage in volts (120 V used in calculations), Eff the efficiency as a decimal and pf the power factor as a decimal.

One HP equals 746 W. If a motor is rated at a true HP, then it delivers 746 W of mechanical power. Single-phase motors are never 100% efficient in converting electrical energy into mechanical energy, so the amount of electrical power consumed by a motor is considerably higher than the mechanical power delivered. In fact, losses from heat and friction mean that a typical single-phase motor is (at best) approximately 60–70% efficient. A figure closer to 60% is more realistic for the small induction motors generally found in many power tools. Hence, a genuine HP motor requires ≥1,250 W of electrical power to deliver its rated power. In short, a 1-HP motor will need ≥10 A of current at 125 V or 5+ A at 250 V to realistically deliver a true 1 HP from the same motor. This is a good rule of thumb.

The power factor of an alternating current (AC) electric power system is defined as the ratio of the active (true or real) power to the apparent power where (a) active (true or real) power is measured in watts and the power drawn by the electrical resistance of a system doing useful work, and (b) apparent power is measured in volt-amperes and is the voltage on an AC system multiplied by all the current that flows in. Table 6.7 shows typical values for motors with different HP.

Table 6.7: Power calculation data for motors.

hp	Eff	pf at full load	Quantity	kw (SP)	Amps (SP)	kw (3P)	Amps (3P)
2	79%	0.84	2	3.7	19	2.1	11
5	84%	0.84	1	4.4	44	2.5	26
15	86%	0.86	1	13.0	126	7.5	73

Power consumption for 20 lights will be 20 × 100 W = 2,000 W = 2.0 kW.

Therefore, power consumed by the system containing 22 single-phase motors and 20 × 100 W lights = 23.1 kW. If this system is run for 8 h, the electrical energy consumed will be 8 × 23.1 = 184.8 W h. The power costs will be based on this number.

6.5.1 Basic example as a guideline

A small factory has one 2-HP motor, one 5-HP motor and 20–100 W lights.

Then, power consumption = $(2 \times 746) + (5 \times 746) + (20 \times 200)$ W
$$= 1,492 + 3,730 + 2,000 \text{ W}$$
$$= 7,222 \text{ W}/1,000 = 7.222 \text{ kW}$$

Power consumption if all work for 8 h:

$$= 7.222 \times 8 = 57.776 \text{ kW h}$$

If all motors are started at the same time, add 50% additional load in watts.
Now, $W = V \times A \times$ power factor

$$7,222 = 440 \times A \times 0.8$$

Therefore, amperes = 20 51 per phase.

A three-phase power of 440 V to 50 Hz wiring system to suit 30 A/phase and a three-phase 440 V circuit breaker trip switch with individual start/stop switches at point of power entry should be provided as a safety measure. Installation of a standby generator will greatly help in minimising downtime due to power failures.

6.6 Calculating water requirements

The need for water is another essential utility for any manufacturing operation. In this case, water is needed for production formulations. Other common needs are for the washing of machine mixing cylinders, emergency showers, eyewash stations, offices, recreation rooms, sprinkler systems as well as other safety systems and other applications. A small-volume foam producer may opt to have his/her water source from the mains or from ground wells. A large-volume foam producer requiring large volumes of water will probably opt for a water supply from a main source. If the factory is located in an industrial zone, the water supply will be available from a main central source delivered to the factory location. In any case, a tank and pumping system has to be put in place. Once the volume of water required is calculated, the sizes of the tanks can be decided. This can be an overhead or ground-level system with water-level indicators and automatic pumping to maintain minimum requirement levels.

The following data will be helpful in determining the sizes of tanks required.

For square or rectangular tanks, $L \times W \times D$ = volume
where L represents the length (ft), W the width (ft) and D the depth (ft).
Therefore, volume in $ft^3 \times 7.47$ = gallons.

For round tanks, $3.1417 \times R^2 \times D$ = volume

where R represents the radius =1/2 diameter, and D the depth.
Then, volume in ft^3 × 7.47 = gallons.

Water quality may or may not be important depending on the end requirement. For example, drinking water will have to be good quality, whereas water for washing purposes may not require special qualities. Depending on the location and supply, water may have to be treated to reach the desired standard. High calcium content is generally a common problem and if not treated will create problems (especially for machines). Industrial water supply systems may require additional attention by way of cooling or heating, and these devices can be incorporated readily into any system.

6.7 Compressed air

In the manufacture of PUR foam, compressed air is an important and required utility. An entrepreneur or a small-volume foam producer may be able to operate with one to two compressors, whereas large-volume foam producers need a steady, large-volume supply. This is generally achieved by an 'air compressor' room in which several air compressors are connected to an accumulator tank from which the compressed air is drawn through different lines to supply whatever operation that needs it. In the manufacture of flexible PUR foam, the three basic needs are pneumatic operation, foaming and vacuum creation.

Air flows through pipes and valves readily, quickly filling spaces. It can be compressed to high pressures and stored as potential energy. If this energy is tapped, it can be used for many types of work. Compressed air is so versatile that it can be considered to be an essential utility in any industrial operation. However, compressed air is not very efficient in that for every 5 HP of energy that is put into compressing air, only 1 HP is available as potential energy. Compressing air produces a stored volume at pressure and the byproduct of this process is heat, which is wasted. A wide variety of air compressors are available from many manufacturers, and a purchaser will have a wide choice to select the most economical ones.

Compressing air creates heat and concentrates water and other contaminates. These must be removed for efficient use in industrial air systems. Delivery lines will have water traps, valves and other devices to ensure that pure dry air is delivered for efficient operations.

6.7.1 Air pressure versus volume calculator

In general, any foam factory will have many machines and equipment that will need air for operations. The two main aspects of air supply will be sufficient air pressure and the total volume of air. Provision will have to be made on the assumption that all pneumatically operated machines may be working simultaneously.

Example: for use as guideline:

Existing maximum pressure	102 psi	7.0 bar		
Required pressure	141 psi	9.7 bar		
Existing compressor capacity	573 cfm	270.5 L/s	16.2 m^3/m	972 m^3/h
Additional capacity required	191.9 cfm	90.5 L/s	5.4 m^3/m	324 m^3/h
Total required capacity	765 cfm	361.0 L/s	21.7 m^3/m	1,302 m^3/h

Therefore, the required increase in air volume = 33.5%.

6.8 Financial indicators

When setting up a manufacturing operation for making flexible PUR foams (or indeed any other business), an investor or an entrepreneur will want to work out the feasibility of such a project. In the case of flexible PUR foams, it is a very viable venture due to well-established and expanding market demands. Nevertheless, a manufacturer would want to know which products to select for manufacture in relation to the capital investment. Financial ratios as given below greatly help in decision making and, if used as performance indicators after manufacturing has started, will show how well a business is doing.

6.8.1 Breakeven point

A breakeven analysis will indicate how much sales volume is needed to meet total expenses, with all sales beyond this point being considered to be profit. The break-even is especially useful in product pricing policy and also to determine what percentage of the market that should be targeted.

Use the formula, BEP = TFC/(USP – VCUP)

where BEP is the breakeven point (number of units), TFC the total fixed costs, USP the unit selling price and VCUP is the variable cost per unit product.

Fixed costs are costs that have to be paid irrespective of whether products are made or not. Costs include rent, indirect labour and telephony. Variable costs are costs that vary directly with production, such as material costs, direct labour and overtime. In the case of a foam manufacturer, because it is very likely that more than one product will be made, a simple BEP unit can be taken as a kilogram or metric ton of material.

Example: A foam manufacturer's fixed costs for producing 100,000 foam cushions is $30,000 per year. Variable cost per unit is $7.00, made up of $2.20 for material, $4.00 labour and $0.80 overheads. The selling price is $12.00 per cushion.

Then, applying formula, BEP = 30,000/(12.00 − 7.00) = 6,000 units.

The foam manufacturer will have to make and sell 6,000 cushions before starting to make a profit.

Another Example

A foam manufacturer makes foam cushions. Each cushion retails at $5.00. It costs $2.00 to make each one and the fixed costs for the period are $750.00. What is the breakeven point in units and in sales revenue?

Applying the formula, X (units) = FC/(SP − VC)

$$= \$750/(\$5 - \$2) = \$750/\$3 = 250 \text{ cushions.}$$

$$\text{Sales volume} = \text{breakeven units} \times \text{SP} = 250 \times 5 = \$ 1{,}250.$$

6.8.2 Return on equity

The return on equity (ROE) or the return on investment (ROI) is the 'bottom line' measure for investors. This measures the profits earned for each dollar invested in a business ROE or ROI is calculated as

ROE = net income ÷ equity

6.8.3 Return on assets

The return on assets is an indication of how effectively a business's assets are being used to generate profits. It is calculated as

return on assets = net income ÷ total assets

6.8.4 Gross profit margin

The gross profit margin (GPM) forecasted in a feasibility analysis will project the gross profit possibilities for paying off bank loans, bank interest and partner's drawings from the net profit possible. The GPM is a measure of gross profit earned by a business on its sales during a specific period and can be expressed as a percentage.

Calculation:

$$\text{gross profit} = \{(\text{sales} - \text{cost of goods sold}) \div \text{sales}\} \times 100$$

6.8.5 Cash flow

Cash flow is a very important and critical area for any business. This is an exercise with cash availability versus expenses or disbursements. That is, investments, bank loans, bank facilities and sales as income versus money needed for payments. This will show how well a business has been planned, with sufficient initial capital, sales and other activities, to generate sufficient funds on a positive accrual basis and whether more cash is needed at any point in the future or not.

6.8.6 Contribution margin

In a proposed multiple-product manufacturing operation or in an existing one to maximise profits, it may be essential to decide which products are viable to make and which should be eliminated. To do this, a foam manufacturer may use the 'contribution' factor for each product he/she makes. A contribution margin is the amount of money a product generates towards meeting overheads and is calculated as unit sale price minus the unit variable costs of that product. Some products may give better contribution margins than others but may have a limited or small market volume. Conversely, smaller-contribution products may have very large market volumes. In decision making, all these factors have to be taken into account to arrive at the best combination.

6.8.7 Debt to equity ratio

If a business applies for a loan or a line of credit, the debt to equity ratio is probably the first ratio a bank will look at. To calculate the debt to equity ratio, the following formula is used:

Total liabilities/owner's equity = ratio >2.0 (anything less will be considered risky)

Note: The above financial indicators can be used as guidelines, and standards can be set as per current business financial practices.

6.9 Quick performance indicators

Instead of waiting until the periodic financial statements are made available to management to study the ongoing performance of an operation, there are certain indicators management can use to assess its performance and take corrective action, if necessary. For a foam manufacturing operation, the indicators described below could be very useful.

6.9.1 Return per kilogram

The return per kilogram is the total gross value of sales and saleable goods versus raw material consumed during a given period of time, for example,

1,000 foam mattresses @ $115/each = $115,000.
Therefore, the return per kilogram   = ($115,000 ÷ 1200) = $95.83 per kg.

Then, compare with the pre-set standard.

6.9.2 Cost per kilogram

Total gross costs of operation versus cost of raw material consumed during a given period of time, for example,

Total costs = ex-factory + marketing + administration + other = $72,000
Raw material consumed during this same period = 1,200 kg
Therefore cost per kg = ($72,000 ÷ 1,200) = $ 60.00 per kg.

Compare with the pre-set standards.

6.9.3 Breakeven analysis

The breakeven analysis is the study of how much sales will be needed to meet total costs. In the case of a foam manufacturer, the total amount of kilograms to be produced and sold would be a good guiding factor. If sales are seasonal, this indicator can be worked out accordingly. Carry out this exercise as discussed earlier.

6.10 Recommended basic foam formulations

Table 6.8 shows some practical basic formulations for foam manufacture. These are formulations that have been used in general practice but slight adjustments may be required depending on the operating temperatures and methods used for mixing.

Table 6.8: Basic formulations for flexible foams.

Raw materials	kg				
Polyol (MW 3,000)	100	100	100	100	100
TDI 80/20	55	50	47	42	38
Methylene chloride	6	5	3.6		
Water	3.6	3.2	2.98	2.57	2.2
Silicone surfactant	1.2	1.0	0.92	0.82	0.72
Tin catalyst	0.25	0.23	0.21	0.20	
	0.21				
Amine catalyst	0.23	0.24	0.24	0.27	0.27
Colour	0.38	0.35	0.30	0.28	0.25
Process temperature 22–25 °C					
Density	18	21	23	28	32

Note: These formulations are provided in good faith and no responsibility is undertaken by the author to say they will work successfully under different conditions. It is best to consider them as guidelines for formulating. A prior laboratory test will greatly assist a foam manufacturer. The cost calculator in Table 6.9 will help to work out the estimated cost of a foam operation before actual production.

Table 6.9: Cost calculation template – example for 23 density.

Raw material	kg (a)	Cost/kg (b)	Total cost (a × b)
Polyol	100	–	–
TDI	47	–	–
Methylene chloride	3.6	–	–
Water	2.98	–	–
Silicone	0.92	–	–
Tin	0.21	–	–
Amine	0.24	–	–
Colour	0.30	–	–
Total weight	155.25	Total cost	X
Weight loss (gas) 5%	7.76	–	–
Net weight	147.49	–	–
Less skins 10%	14.75	–	–
Good foam weight	132.74	–	–
Cost/kg	–	–	X/132.74

Note: This is only an example. Gas losses and skin losses will depend on the processing parameters and methodology. The skin wastes can be recycled and sold, giving a 'kickback' towards income/ profits.

6.11 To save waste on foam buns (blocks)

On continuous foaming lines when buns are cut to equal standard length there will be a lot of waste during conversion to cushions and mattresses. It is recommended that these buns are cut to lengths in such a way to match orders in hand and the operators on the foaming line should be instructed on the exact sizes required.

7 Plant machinery and equipment

7.1 Mould design

The manufacture of flexible polyurethane foam (FPF) requires moulds from simple designs to the more complex ones for continuous foaming and for moulded foam. Computer-assisted mould design and mould-making are the result of microprocessor technology, and other advanced technologies are also available. Computer-aided designing and computer-assisted machining (CAD/CAM) can be used to design the more complicated moulds required for moulded foam products compared with the large moulds required for the production of foam blocks. CAD/CAM systems are especially useful in designing contoured or multi-cavity moulds.

During the early stages of industry, most moulds had to be more or less handmade, so it was a highly labour-intensive and costly process. Modern-day technologies have allowed the industry not only to manufacture moulds at reasonable costs but also with precise tolerances, to minimise waste, increase moulding cycle times as well as short completion and delivery times. In general, a mould design starts with a sketch, drawing or an actual sample of a product. In large foam block manufacture, moulds are designed by predetermined dimensions of the required final product.

For moulded foam products, one may opt to design and make a pilot mould to turn out prototype products before investing in a permanent one. Moulds for plastics have a lifespan, that is, they are most effective for precision products around a minimum of 20,000 to 30,000 mouldings. Some may last longer depending on the skills applied by a mould maker, and the type of product being made also has bearing in a lifespan of the mould. Some of the important parameters for designing and making a mould are: product geometry, weight, shrinkage, ventilation, closing/opening system, engraving, the plating for the required finish (e.g., smooth, rough, matt or glossy surface), gas ventilation and the type of material being moulded. Some materials that can be used for prototypes include wood, gypsum plasters, fibre glass, light metals and plastics.

When designing a product, one must consider the interaction of three major components: (1) material, (2) design and (3) manufacturing. The correct selection of material is often the most involved, complex and certainly the most costly aspect of product design, which often requires development, testing and evaluation. Factors for material selection must take into consideration temperature effects, fabrication time, aging of material and the ultimate cost to the customer.

https://doi.org/10.1515/9783110643183-007

7.1.1 Design philosophy

The goal of a systematic design process is to control the various stages that influence the finished product and to understand the interactions between those stages. Typically, a product design involves the following:
- Marketing
- Aesthetic or industrial value
- Engineering design
- Tooling
- Manufacturing
- Delivery period

Traditionally, mould-making was handled by 1 or 2 people, and the methods and technology available to them resulted in trial and error, which would result in high labour costs and delays. With modern-day high-tech methods and concurrent engineering, these factors are eliminated. One of the main advantages of concurrent engineering is that the whole project is done as a team effort with constant communication and interaction between all team members from marketing to manufacture. Although the functional and aesthetic design of a part must be finalised before the engineering design and calculations begin, other aspects such as material choice and manufacturing process can be done concurrent with industrial design.

An engineer can often draw on the experience of other individuals if designing a new product or reproducing the product from a sample. For example, product databanks are available that provide information on basic design requirements, materials selection and other design-dependent characteristics. The engineering design aspect of a product can be broken down into main stages as follows:
- Define product requirements
- Preliminary CAD model
- Material selection
- Mould operation requirements
- Design for manufacturing
- Make a prototype
- Check suitability (adjustments where necessary)
- Final sample
- Customer approval

7.1.2 Defining product requirements

The end-use requirements of a product are decided during design. At this point, it is preferable to specify quantitative requirements such as maximum loads and operating temperatures. In addition to product dimensions, industrial design aspects and economic issues, a designer must consider the loading and environmental conditions, as well as regulatory and standard restrictions.

7.1.3 Loading conditions

The type, magnitude and duration of loading must be determined. These loadings can occur during processing such as the stresses and deformations that may occur during mould filling, demoulding, assembly, packing, shipping and the life of a product.

7.1.4 Environmental conditions

These can include exposure to sunlight and consequently ultraviolet (UV) rays, humidity or chemical environments. Chemical environmental factors can be the use of lubricants, cleaning agents or other materials that may lead to the chemical degradation or environmental stress cracking of a mould. Special attention is needed if corrosive materials are being moulded and if sufficient data are not available. The chemical resistance of the materials must be determined by testing.

7.1.5 Dimensional requirements

The load, functionality, shrinkage, tolerances, as well as the mould design decide the size and thickness of a moulded product. However, cost, manufacturability, material choices and polymer physics bring several constraints at this stage of design. Material choice and polymer physics have a profound effect on the design, manufacturing and performance of a product.

7.1.6 Preliminary computer-aided design model

In making moulds for PUR, the general requirements for moulds are for intermittent foaming in which foam blocks are made one by one and for moulded foam products. In the continuous foaming process, the 'mould' is the moving conveyor, with horizontal or vertical operation, and the concept of mould setup is quite another matter.

Once the load, strength and functional requirements have been calculated and preliminary decisions made, together with the industrial design team, the mould designer can generate a three-dimensional (3D) CAD model that will help visualise the design and its functionality and point to possible problems. This model can then be modified and subsequently used to generate a finite element model to produce a prototype.

7.1.7 Material selection

For most plastic products, material costs account for >50% of the manufacturing expense. Once all end functions and requirements for the product have been specified, the mould designer can start searching for the best available and most suitable material. During this process, the designer must take into consideration material properties such as creep, heat distortion, impact strength and durability to name a few. Databanks can provide already tested and tried standard results with tolerance limits.

7.1.8 Process selection

Ultimately, product quantity as well as the geometry and size of the mould may decide which process should be used to manufacture a specific product. Alternatively, if a mould is required for an existing machine or process, then the mould(s) must be designed to accommodate them. The moulding conditions and processes for PUR foam are different to standard manufacture of plastic parts because chemical reactions must be considered. For example, the manufacture of solid or hollow plastic parts may involve injection moulding, blow moulding or compression moulding, whereas production of PUR foam involves a 'pour', pouring or filling a mould.

7.1.9 Process influences

The mechanical and physical properties and dimensional stability of moulded polymers are strongly dependent on the anisotropy of the finished part. The structure of a final product is influenced by the design of the mould cavity (e.g., the type and position of the pour inlet, clamping system, venting, heating and cure time). After moulding or shaping of a product, it must set or cure by cooling or cross-linking. A thermoplastic material will harden as the temperature is lowered, either the melting temperature for a semi-crystalline polymer or the glass transition temperature (T_g) for an amorphous thermoplastic. A thermoplastic can soften again as the temperature of the material is raised above the solidification temperature. For PUR that are thermostats, cooling leads to cross-linking of molecules upon solidification. The effects of cross-linkage are irreversible and lead to a network that hinders the free movement of the polymer chains independent of the material temperature. Solidification leads to residual stresses and consequently to possible warping, perhaps one of the biggest problems facing a design engineer.

7.1.10 Orientation of the final product

During processing, the molecules, fillers and fibres are oriented in the flow and may greatly affect the properties of the final product. When thermoplastic components are manufactured, the polymer molecules become oriented. The molecular orientation is induced by the deformation of the polymer melt during processing. The flexible molecular chains become stretched and, because of their entanglement, they cannot relax fast enough before the part cools and solidifies. At lower processing temperatures, this phenomenon is multiplied, leading to an even greater degree of molecular orientation.

During the manufacture of thermosets (PUR), there is no molecular orientation because of the cross-linking that occurs during the solidification or curing reaction. A thermoset polymer solidifies as it undergoes an exothermic (heat-giving) reaction and forms a tight network of interconnected molecules. The properties of the final product may depend to a large extent on the fillers and additives incorporated into the foam, especially factors such as density, shrinkage and warping.

7.1.11 Computer simulation

Computer simulation of polymer processes offer the tremendous advantage of enabling designers and engineers to consider virtually any geometric and processing option without having to incur the expense of making a prototype mould. The ability to try new designs or concepts gives the design engineer the opportunity to detect and fix problems before proceeding to a prototype mould and production. Additionally, the process engineer can determine the sensitivity and limitations of processing parameters to obtain a desired moulded foam product. For example, computer-aided engineering offers the mould designer the flexibility to determine the effects of different positions of pour-holes, cooling systems and moulding cycles.

7.1.12 Shrinkage and warping

Shrinkage and warping are directly related to residual stresses, which result from locally varying strain fields that occur during the curing or solidification stage of a manufacturing process. Such strain gradients are caused by numerous thermomechanical properties and temperature variation inside the mould cavity. Shrinkage due to cure can also have a dominant role in the development of residual stress in thermosetting polymers.

Minimising warping is one of the biggest challenges for a mould design engineer. This can sometimes be achieved by changing a formulation in a foam mix. In the processing of PUR foams, the interactions of the reactants in a mix have a major role, but the temperature parameters inside and outside the mould are very important.

7.1.13 Basic moulds for flexible foams

Flexible foams can be made in different ways depending on the end applications. For example, they can involve simple hand-mixing and pouring into wooden moulds; sophisticated low-pressure or high-pressure dispensing machines; large-volume and advanced continuous foaming systems; and integral skin flexible foams where machines with dispensing heads or mixing heads are used with single-cavity or multi-cavity moulds.

7.2 Foaming machinery and equipment

Since its inception, PUR foaming has undergone tremendous change from the humble electric drill method for the very small foam producer to the modern-day semi-automatic and fully automatic foaming systems. Foaming systems take the form of making foam blocks, one by one, or in a continuous form and then cut into 'buns'. While horizontal continuous foaming is the most popular, vertical systems are also available. For the foam, small- or large-part moulder, there are two basic types of machines: (1) low-pressure and (2) high-pressure metering/dispensing machines.

7.2.1 Low-pressure machines

Low-pressure machines are most often selected for low-volume production applications. Mixing of the chemicals takes place in a chamber, which includes an electric motor–driven or hydraulically operated dynamic mixer resembling an auger with multiple grooves. The dynamic mixer can have a speed range from 3,000 rpm to 8,000 rpm depending on the formulation used. Piston-type chemical injectors pneumatically retract to allow the blending of multiple chemical streams at pressures ranging from 30 psi to 500 psi with mixing by the dynamic mixer. When the required 'shot' of blended chemicals is dispensed, the injectors close. After a desired production run, the mixer is usually flushed with solvent and/or water and compressed air to purge remaining chemicals from the mixing chamber.

7.2.2 High-pressure machines

High-pressure machines are used predominantly in applications that require measured shots to produce moulded foams. The principle involves the impingement blending of chemical components under high pressure in the mixing chamber of the mixhead. A hydraulically controlled plunger in the mixhead retracts, allowing two or more streams of chemicals to merge together under high pressure in the range of 1,200 psi to 3,000 psi in the chamber. After metering, and a precise shot of chemicals, the plunger closes and cleans out the mixhead. This system is self-cleaning and does not require flushing.

7.2.3 Images of foaming machines

There are many different types and systems of foaming machines available. The ones described below are commonly used by most manufacturers of FPF.

Figure 7.1 shows a large foam block made using a batch foaming machine. Figure 7.2 shows a typical self-contained batch foaming system. Continuous foaming systems are shown in Figures 7.3 and 7.4. Large quantities of foam waste are a part of any foam-manufacturing operation, so a foam waste re-bonding machine is essential (Figure 7.5).

Batch foaming machine (BFP201A)
www.foam-machinery.com

Figure 7.1: Flexible polyurethane foam block. Reproduced with permission from Modern Enterprises, India.

7.3 Foam cutting and fabrication

Once the foam blocks have been made and stored (after a minimum curing period of 24 h), these blocks can be taken out on a first in–first out basis, which ensures that

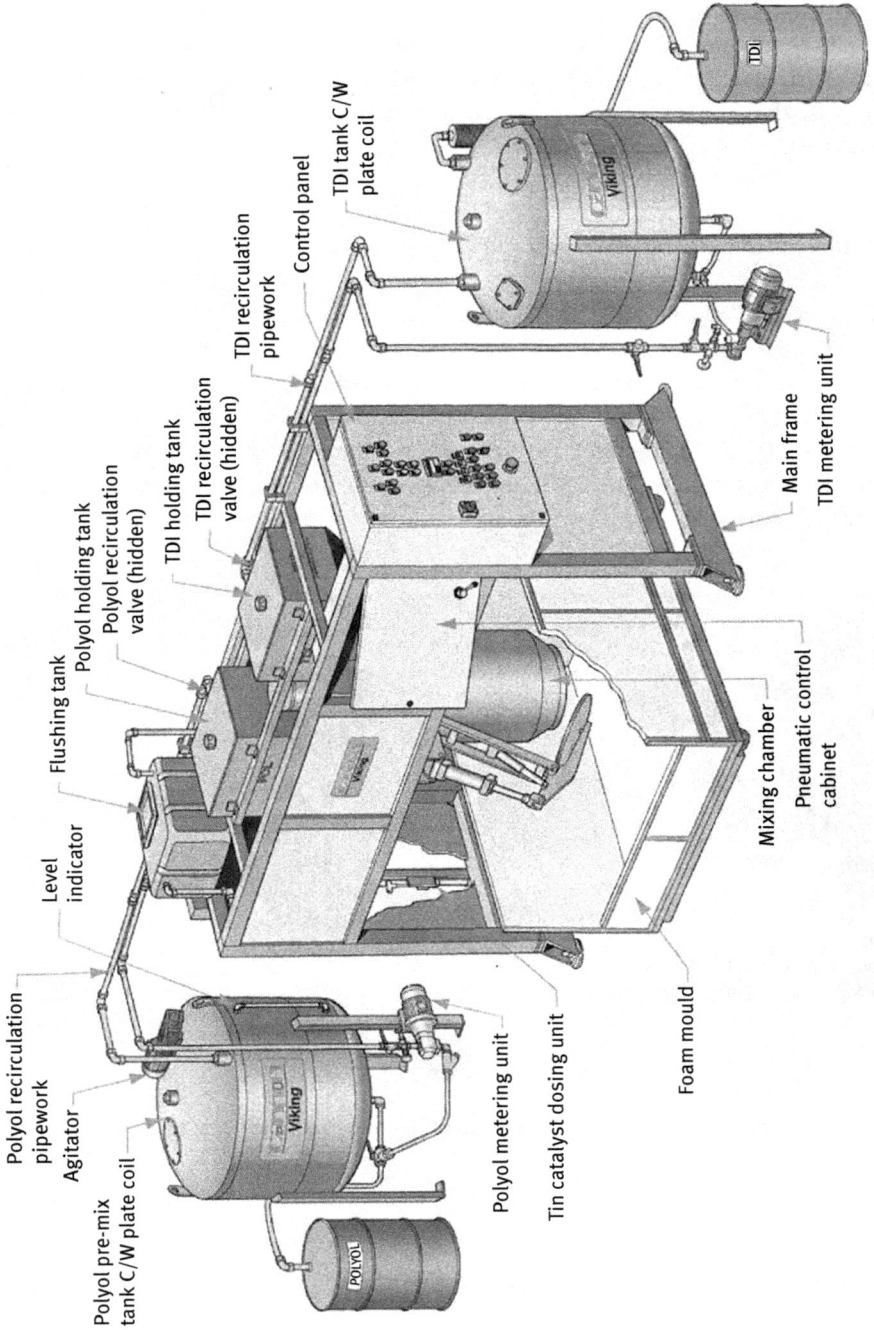

Figure 7.2: Batch foaming system. Reproduced with permission from Canon Viking, Manchester, UK.

Figure 7.3: Continuous foaming process. Reproduced with permission from AS Enterprises, India.

Figure 7.4: Continuous foaming machine. Reproduced with permission from Modern Enterprises, India.

appropriately cured blocks are taken out in sequence. These foam blocks can be converted by cutting and fabrication into final products such as slabs, sheets, profiled items, mattresses, cushions, mattress toppers, items for furniture, automotive applications, railway wagons and boats, pillows, temperature, vibration- and noise-damping materials, packaging, carpet underlay, car and bath sponges, domestic sponges, sanitary pads, clothing items, shoulder pads and textile lining, backing for leather and polyvinyl chloride handbags, suitcases and shoe linings, toys,

Figure 7.5: Foam waste re-bonding machine. Reproduced with permission from Modern Enterprises, India.

children's day care items and medical applications. Standard and specialised cutting and fabricating machines are available for conversion of foam into all of these products.

When planning a foam-cutting and fabricating department, it should accommodate all the required machines plus space for adding more machines later on, space for blocks, as well as space for holding temporarily the finished products until quality-control checking and removal to the finished goods store. In general, blocks are transported from the curing store to the conversion area by conveyor, forklift truck or manual trolleys depending on the size of the operation. The first action would be to trim all sides of a block ('skins') to arrive at the overall size of the 'virgin' foam block desired. For a foam manufacturer, it is advantageous to invest in good-quality machines such as semi-automatic or automatic electronically operated machines that will give fast and very accurate conversions, even though the initial costs will be high. These investments are wise in the long run because high and steady outputs with minimal scrap losses compensate the extra costs.

The most rational way of trimming the blocks (i.e., removing the side skins and simultaneously obtaining a fixed width) is by moving the blocks on a conveyor with adjustable vertical cutting on both sides. Thus, the cutting/trimming operation is on both sides during transport of the blocks for further conversion (e.g., stack slitting), thereby saving time, handling and labour. The foam blocks can also be trimmed one side at a time by a manual vertical cutting machine. This was the traditional way, but was awkward, slow and labour-intensive as compared with modern electronic machines.

7.3.1 Horizontal cutting

After the blocks have been trimmed, they can be split into sheets, toppers and mattresses to predetermined thicknesses using horizontal cutting machines. There are different types of these machines as described below:
- The conventional slitting machine cuts sheets one by one (i.e., each sheet must be removed after cutting before the next sheet can be cut). This type of machine has a rugged knife-gib and is available with or without automatic controls.
- A stack slitting machine slits a block (or blocks) into a stack of sheets. That is, the sheets can be left on the machine until a part or the entire block has been converted into sheets. This type of machine has a thin knife-guide and is usually operated automatically. There are also a number of specialised slitters available for specific purposes and different materials.

7.3.2 Vertical cutting

Vertical cutting machines with band knives are used for all types of cutting work and are the most frequently used type of machines because of their versatility. Machines are available in fully automatic, semi-automatic or manual versions and with different dimensions and equipment to meet most requirements. Typical products made on these machines are: cushions and sheets for furniture, packing, sponges, pads, clothing items and similar items.

7.3.3 Specialised cutting

Over the years, several specialised cutting machines have been developed to facilitate the manufacture of the various products created by market demand:
- Bevel cutting machine: for furniture, insulation items, profiles and padding.
- Peeling machines: for cutting foam into very thin sheets in roll form for quilting in mattresses, laminating with textiles, packing material or carpet underlay.
- Profile cutting machines: for special mattresses and furniture needs, packing.
- Edge profiling machines: for profiling/curving edges of furniture, cushions.
- Contour cutting machines: for cutting complicated items.
- Cross-cutting machines: for cutting sanitary washing pads in stacks of thin sheets.
- Convoluting machines: where a single sheet is cut into 'hill and valley' shapes.
- Punching presses: for toys, packing, insulation.

Cured foam blocks are first trimmed using horizontal and vertical cutting machines to remove the skins. Then, these same machines (shown in Figures 7.6 and 7.7) are used to cut the trimmed blocks into sheets, cushions and slabs. If, however, faster

Figure 7.6: Foam-cutting machine. Reproduced with permission from Modern Enterprises, India.

Figure 7.7: Continuous foam-cutting machine. Reproduced with permission from AS Enterprises, India.

cutting is required because of large volumes, then multiple circular block-cutting machines shown in Figures 7.8–7.10 can be used. All waste to be re-bonded has to be reduced in size, and for this a foam-shredding machine is used (Figure 7.11).

Figure 7.8: Model 1 multiple block-cutting machine. Reproduced with permission from Modern Enterprises, India.

Figure 7.9: Model 2 multiple block-cutting machine. Reproduced with permission from AS Enterprises, India.

Figure 7.10: Model 3 multiple block-cutting machine. Reproduced with permission from Modern Enterprises, India.

Shredding machine with blower
www.foam-machinery.com

Figure 7.11: Foam waste-shredding machine. Reproduced with permission from Modern Enterprises, India.

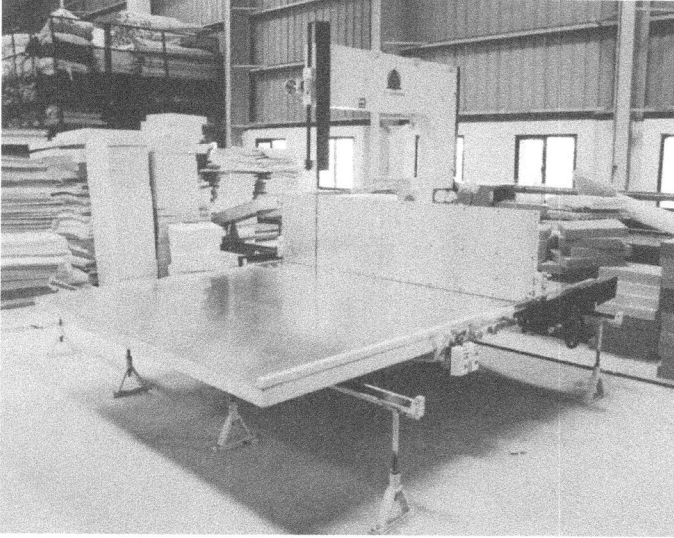

Figure 7.12: Vertical cutting machine. Reproduced with permission from Modern Enterprises, India.

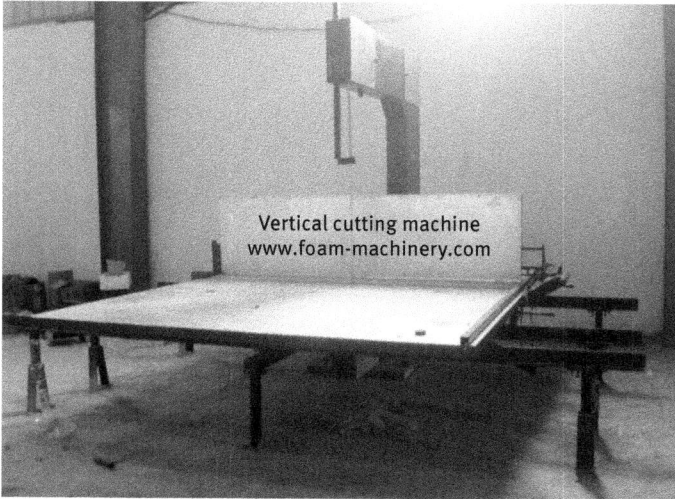

Figure 7.13: Vertical foam-cutting machine. Reproduced with permission from Modern Enterprises, India.

8 Manufacturing processes for flexible foams

Flexible polyurethane (PUR) foams can be made in many ways. The criteria for deciding which process is to use depends on the capital available and more so the targeted volume of production. Good quality can be achieved with any process, but the manufacturer must involve operational ability, wastage and environmental concerns. In general, the machinery suppliers will provide installation, technical and training aspects, whereas waste is a process control and advance planning that will offset any environmental issue.

A large portion of flexible PUR foams is made in blocks by the discontinuous (intermittent) process in simple boxes or in a continuous process known as 'slab stock foaming'. In the former, small or large blocks are made one by one to desired sizes and specifications; in the latter process, foaming is done in a continuously flowing process and the resulting foam is cut into desired lengths of blocks called 'buns'. In each case, the foam blocks are transported to a holding area for post-cure before being taken for cutting/fabrication.

8.1 Box process for small-volume producers

The box process is a simple method for producing good-quality flexible PUR foam blocks in small sizes. This method may suit an entrepreneur with limited capital and space. This is essentially a small-volume operation and recommended for manufacturing foam cushions, pads and other similar products targeted for the furniture industry. This method may be suitable in 'developing' countries, where an entrepreneur with limited resources and technical know-how is targeting a small-volume market. This type of operation could produce ≈2 metric tons of foam per month.

In general, standard foam cushions sizes are 20 × 20 × 4 inches and 24 × 24 × 5 inches. The box mould can be made of wood ≈1-inch thick with detachable sides and the bottom with a groove to accommodate the four-sided box to prevent the liquid foam mix leaking when poured. The dimensions of the box (height, width and length) will depend on the sizes of foam blocks required to be able to cut the desired sizes of cushions. For mixing, a single-phase electric drill of 900–1200 rpm with adjustable speeds and a 30-inch long and 3/8-inch diameter steel shaft should be suitable. To the bottom edge of this shaft a round flat disc of ≈0.4-inch diameter with flanges should be attached firmly. The design or pattern of this disc should be constructed so as to achieve a good homogenous mix. Other basic equipment required would be a small weighing machine, protective wear, a floating lid made of very light wood (e.g. plywood), a cutting machine, a few plastic buckets and small containers. An entrepreneur will have a choice of using raw materials as already mixed two-component

https://doi.org/10.1515/9783110643183-008

systems for ease of processing or purchasing the raw materials as separate ingredients and formulating on his own. In the former case, different systems have to be purchased to obtain different densities and properties; in the latter case, the advantage is that one can formulate any combination of ingredients to get the desired end results.

8.1.1 Process

The mould is set up on the base and is lined with a thin polyethylene film, or the mould sides are sprayed/waxed with a release agent. All chemicals are weighed separately and mixed together in a large mixing vessel or plastic bucket except toluene diisocyanate (TDI). Finally, the TDI is introduced into the polyol mix and then mixed quickly and within a few seconds poured into the mould. The mix must be poured into the mould in a liquid state because delay will result in the mix beginning to 'cream' and the batch will have to be discarded.

The liquid mix at the bottom of the mould will start to 'cream' and rise slowly. When the foam has risen about one-third of the mould height, the floating lid is placed gently on the surface of the rising foam to prevent a meniscus (round surface) being formed, which will result in waste. If the floating lid is introduced appropriately, the foam block will have a flat top. Cooling and setting time should be ≈5–10 min and then the foam block can be de-moulded. The process can be repeated as many times as desired. The resulting foam blocks are then transported to a holding or post-curing area with good ventilation and stored for 24 h to allow the foam to cure further by cooling, and to complete the emission of gases because of the exothermic reaction taking place. The blocks must be placed ≥1-foot apart and without stacking one on top of the other to prevent fire hazards. After 24 hours, the blocks are ready for cutting or fabrication.

8.1.2 Advantages

The following reasons should be considered seriously, especially by an entrepreneur:
- Low capital investment
- Small operating space
- Only 2–3 operators required
- Low costs/high profit margins
- Only single-phase power required
- All required equipment can be fabricated
- Low inventory costs
- Minimal environmental issues
- Choice of two-component raw material systems for easy processing

8.1.3 Disadvantages

The following reasons should also be taken into consideration by an entrepreneur:
- Low-volume production
- Inability to avail price advantages because of bulk purchases
- May encounter difficulty in small purchases of raw materials

Note: This process is presented as a real project in detail later in this chapter.

8.2 Discontinuous process

The discontinuous (intermittent) process involves large blocks of flexible PUR foam of general standard size $2 \times 1 \times 1$ m are made one by one. This method is suitable for a foam producer making $\geq$100–150 metric tons per year. A producer has the choice of using a manual, semi-automatic or automatic operation. Many types of machinery with standard or advanced mixing or dispensing heads are available but the operating principles are more or less identical.

8.2.1 Manual process

All required raw materials can be purchased in bulk in drums, totes or smaller containers. In this operation, it is advantageous to purchase the polyol and isocyanate in drums for easy handling and to avoid totes because of the difficulty in extracting the liquids. Raw materials are also available in two-component systems and will be easy to mix but with the limitations described above. There are two systems for mixing and pouring. A large mould $\approx 2 \times 1 \times 1$ m on wheels or on rails, with all sides lined with thin polyethylene film, waxed or sprayed with a release agent such as silicone oil, is made ready and brought under a large mixing vessel of a foaming machine. This mixing vessel will have a bottom lid opening and closing hydraulically, and could move up and down.

This mixing vessel is centred in the mould and brought down to $\approx$1 foot above the bottom of the mould with the bottom closed. All ingredients are weighed separately and introduced into the mixing vessel except the isocyanate and then mixed for a predetermined period. The isocyanate is put into the mix and mixed for <6 s. The bottom lid of the mixing vessel is activated and opened, and the mixed liquid flows into the bottom of the mould. Simultaneously, the mixing vessel is raised slowly and the resulting foam rises slowly. When it has reached about one-third, its mould height and before the meniscus (curve) starts to form, a light floating lid is placed gently on the rising foam. This results in a flat top foam block, thereby saving a huge curved surface that would otherwise have been considered to be waste.

The mould is then wheeled out and another mould (if available) can be used and the process repeated or the same mould can be used repeatedly. The moulded foam block will take about 10 min to cure and when it is not tacky, can be de-moulded. The mixing vessel must be cleaned or washed with water (or a cleaning agent such as methylene chloride) to remove particles or residues remaining in the vessel before the next cycle. The moulded foam block after de-moulding must be handled with care because it is not fully cured and emits out heat because of the exothermic reaction taking place inside the foam block. The blocks are then transported to a holding or post-cure area well removed from the production area and stored in 1-foot apart and without stacking one on top of each other to counter fire hazards. After 24 h, these blocks can be taken to the cutting and fabrication area.

8.2.2 Advantages

These points will assist in deciding the level of foam production for a foam moulder:
- Reasonable capital investment
- Not labour-intensive
- Raw materials can be purchased on a just-in-time system
- Waste control because of individual blocks
- Blocks can be made with different densities and properties
- Cost-effective production

8.2.3 Disadvantages

The following should also be considered when deciding on a foam operation:
- Inability to get real price advantages with bulk purchases
- Storage of large inventories under safety standards
- Disposal of empty drums/containers requires special care

8.3 Semi-automatic and automatic processes

Chemical ingredients – polyol and isocyanate are stored in large tanks, whereas others are stored in smaller containers. All streams are connected *via* liquid flow systems such as flowmeters, control valves, gauges and pumps to a main mixing/dispensing head. In the case of the semi-automatic operation, all ingredients can be metered according to a predetermined formulation into the mixing head and dispensed into a pre-set mould as described above. The chemical formula settings can be changed for each block or in batches. Hence, to make blocks with different densities/properties. The process is the same as described above and pre-set timers on the foaming machine control the mixing time and lifting of the mixing vessel. After each foaming cycle, the mixing head must be flushed manually by bringing on a cleaning agent by operating a valve/switch before the beginning of the next cycle.

In the automatic operation, a predetermined formula of all ingredients is programmed into the foaming machine. The machine follows the pre-set sequence of bringing in of the metered quantities of all ingredients to the main mixing/dispensing head and mixes them according to the pre-set timings. Finally, the isocyanate enters the polyol mix and is mixed and dispensed into the pre-set mould. All operations are automatic and the procedure is identical. The mixing head is flushed/cleaned automatically, and is ready for the next cycle according to pre-set parameters.

8.3.1 Advantages

Important areas to consider when deciding between manual and machine operation:
- Ease of operation
- Fast production cycles
- Accurate weighing/metering by machine
- Not labour-intensive

8.3.2 Disadvantages

These would help a foam moulder to work out the feasibility of such an operation:
- Large capital investment
- Effective maintenance requirement (especially chemical pumps)
- Maintaining the chemical tanks at an appropriate temperature
- Faulty metering because of malfunctioning pumps will result in foam block waste
- Large holding/curing areas required to accommodate large volume of blocks

8.4 Viscoelastic discontinuous block foaming

With rapid development of viscoelastic ('memory') foam and a large demand in market growth, the production of viscoelastic foam has been growing into large volumes. Over the years, in particular, the mattress market has changed to viscoelastic foam.

The manufacture of viscoelastic foam and production sequences is, in principle, more or less the same as for conventional flexible PUR foams, although speciality grades of polyols and some others are used. Usually, the popular method of manufacture is the single block production in which the liquids are mixed and poured manually or with semi-automatic or automatic foaming systems. Because of the very heavy nature of the foamed blocks, the height of a block is restricted to ≈28–30 inches instead of the standard 42–45 inches for flexible foams, whereas a width of 84 inches can be maintained. Viscoelastic foam blocks require longer final post-curing periods, ≥48 h as compared with the 24 h for conventional flexible foam. During formulating and weighing of the different components, accuracy is very important to avoid distortion of the moulded foam block and shrinkage of the width and length.

The use of the specially designed Blockmatic single machine (made by Cannon Viking) or a similar machine is one method for producing viscoelastic foams in an economical way. An improved version of this model is also with a 150-kg shot/pour with multiple separate metering lines and tank storage for the special polyols and additives used. The shot or pour volume, in relation to the length and width of the block to be produced, will decide the height of the block.

8.5 Continuous process

Manufacturers of >1,000 metric tons per year of flexible PUR foams will opt for the continuous process (also known as 'slab stock' production). In this process, the two main raw materials – polyol and TDI – can be purchased in tankers and pumped directly into large holding tanks. The other components, such as water, auxiliary blowing agents, surfactants, catalysts and other additives are filled into smaller tanks or containers. All these holding sources are connected to a mixing station on a continuous foaming machine by a controlled metering system.

The foaming machine consists of an electronically controlled foaming station, a mixing and dispensing head connected to all the component holding tanks, a moving conveyor with variable speeds, a trough arrangement lined with thick paper to act as the mould for the foam and a movable vertical foam cutting saw mounted on the machine towards the end, a few metres away from the dispensing station. A predetermined formula is programmed into the feed-end directly connected to the special mixing/dispensing head. When activated, all the components are brought on stream in a pre-planned sequence and the reactants in a homogenous mix (still in liquid form) are released onto the bottom of a trough lined with paper. The height between the bottom of the dispensing unit to the bottom of the trough and the conveyor speed is matched to allow the liquid to 'cream' and then rise slowly while moving. The length of the conveyor/trough is designed so that by the time the fully risen foam reaches the vertical cutting area, it has cured sufficiently without being tacky and the vertical cutting saw slices the slab stock into blocks/buns of desired lengths. Machines in which there are no suitable top-flattening arrangements produce slab stock with a curved top, which will be waste. Some manufacturers use a block weighing system after cut off for statistical and quality-control purposes. These blocks are then transported to well-ventilated and spacious holding/post-curing areas and stored 1-foot apart and without stacking one on top of each other to counter fire hazards. The blocks can be taken to the cutting and fabrication area after 24 h. Because of the very large volume of blocks, it is best to use a first-in–first-out storing system to ensure fully cured blocks are taken for cutting and ease of movement.

8.5.1 Advantages

The following are the basic advantages of a standard continuous foam production operation:
- Very large-volume production
- Economical and fast production runs
- Production of slab stock foam with different properties on the same run
- Machines need only a few operators
- Machines have waste-reducing devices

8.5.2 Disadvantages

The following are important points to consider when deciding large-volume productions:
- Heavy outlay of capital
- Very large floor space required for factory layout
- High foam waste if flat-top device is not available
- Paper used for mould formation/lining cannot be reused
- Accuracy in formulation/programming to prevent waste
- Inherent waste at the start and end of production run

Note: A case study from an actual operation is discussed later in this chapter.

Some of the widely used continuous foaming systems for production of flexible PUR slab stock are as follows:
- Standard conventional laydown continuous foaming machines
- Maxfoam continuous foaming lines
- Vertifoam vertical continuous foaming system
- Varimax continuous foaming system
- C-Max high-pressure continuous foaming system
- CarDio continuous foaming system
- High-tech viscoelastic foaming system

8.6 Maxfoam system

The Maxfoam system may specifically suit a medium- or large-volume foam producer that does not require a high level of sophistication but needs the capacity to reach production volumes of a continuous production line. This machine is ideally suited for producing flexible foams for applications such as furniture foam and mattresses.

The Maxfoam process is probably the most widely used and popular continuous foaming system in most countries. Many continuous flexible foaming lines are using the basic principles of the Maxfoam process.

8.6.1 Basic principles

The main difference between the Maxfoam process and a conventional conveyor slab stock process is the trough. The trough is designed to cause a delay between mixing the chemicals and depositing the mix on the conveyor. This delay, which is typically about 20 s, allows some pre-foaming to take place in the trough so that a froth instead of a liquid is laid down on the conveyor. The advantage of this configuration is that the transportation surface where the foam expansion occurs (the Maxfoam fallplate) can be much steeper than the conveyor of a conventional machine. The angle of inclination of a conventional foam conveyor is usually <4°, but the average inclination of this fallplate is >10°.

The steeper angle is possible because the froth coming out of the trough has a much greater viscosity and is of lower density than the liquid mix on a conventional machine. These two factors reduce the possibility of under-run, that is, the running of heavier liquid material underneath older, lower-density foam. This steeper angle also allows a shorter machine to be used at lower outputs and conveyor speeds for the same height of block. The density distribution through the foam block is also improved. A further advantage of the Maxfoam machine is the ability to control the shape of the top of the block by adjusting the trough height and full-rise position.

8.6.2 Basic features of the Maxfoam system

Each foaming system has its own unique features and these are highlighted below:
- Production of foams of varying qualities
- Efficient and economical operation
- Easy to use control panel
- Reliable and accurate metering units
- Easy to feed/rewind paper systems
- Automatic self-adjusting recirculation valves

- Polyethylene film-covered multi-trough paper prevents foam residue
- Easy/efficient flushing of manifold and mixer

All chemical streams are metered by highly accurate and efficient pumps and directed to the mixing head port-ring or to the polyol manifold through three-way diverter valves. This arrangement enables the metering pumps to run at normal output conditions in recirculation mode. During foam production, the chemicals are mixed in the mixing head, which consists of a variable-speed multi-pin stirrer closely fitted inside a cylindrical barrel. The speed can be varied by an infinitely variable speed controller. Two fundamental mechanical items are the trough and the fallplate. The chemical mix is fed from the mixing head through two flexible hoses into the bottom of a deep trough. The foam mix is allowed to start reacting and partially expand in the trough before spilling out onto the fallplate section.

The fallplate consists of several hinged sections. The vertical position of the trough and each section of the fallplate is controlled by motorised screw jacks. This arrangement enables the profile shape of the fallplate to be changed to suit different foam formulations. Onto the fallplate to contain the foam from the trough is fed the bottom paper – a continuous sheet of paper from a horizontal unwinding roll at the back of the machine. Side papers are also fed from vertical unwinding rolls at the start to form the sidewalls. The feed tension of the unwinding rolls is controlled by adjustable brakes from the bottom end of the fallplate. The bottom paper runs onto the main conveyor, which is of metal slate construction. The speed of the main conveyor is variable and is critical for the foaming process.

At the end of the main conveyor, where the sidewall papers are removed from the continuous foam block, are two vertical capstan rolls that are driven at the same surface speed as the main conveyor. This ensures that the side papers travel at the same speed as the foam block. The side papers are wound onto rotating rewind rolls. The bottom paper is also removed from the foam block and wound onto a driven rewind roll below the level of the conveyor. A block cut-off knife or saw is situated after the paper take-off stations. The block lengths desired are pre-set and, as it cuts, it moves synchronously with the main conveyor so as to ensure accurate vertical cuts. After the blocks have been cut, they are transported to a final curing area well away from the production area and allowed to cure for 24 h before being taken for cutting/fabrication.

8.7 Vertifoam vertical foaming system

The Vertifoam vertical foaming system produces perfectly rectangular or round foam with minimum skin loss with improved utilisation of chemicals and full-size blocks of foam at relatively low chemical outputs. The Vertifoam system produces foam continuously in a vertical direction instead of the standard horizontal direction, and the machines are available with rectangular or round tunnels. The round blocks are needed for making thin foam sheeting on a peeling machine for quilting in the mattress industry. Unlike in conventional foaming systems, where a producer has to contend with a continuous thick bottom and top skin which is waste, this system will have minimum skin losses because the heavy top and bottom skins do not arise and all four sides produce only very light skins that for many applications will not need trimming. This system will enhance better density and softness/hardness distribution throughout the block, and variations of these parameters will be at a minimum level. Changes in colour and grade are also possible on the same run with minimal waste, and another advantage is that only a few operators are required to run the machine efficiently.

8.8 Varimax continuous system

The Varimax continuous system enables a foam producer to make rectangular flexible foam blocks with varying block widths on the same production run. Producers who already have a standard Maxfoam machine could, with collaboration of the machine suppliers, convert their existing machine to the Varimax system.

8.9 C-Max high-pressure system

The C-Max high-pressure system involves multi-process slab stock technology. It is more or less a combination of producing foam by the Maxfoam and conventional processes based on the low scrap loss of Maxfoam and the versatility of the conventional process. The C-Max system operates as a high-pressure Maxfoam with liquid laydown and uses the Maxfoam fallplate to improve foam block shape, with foams of uniform cell structure and low pinholes, or as a conventional direct pour, producing higher-density foams and speciality grades. To obtain maximum economy of production, these machines are designed with the standard height of 1,200-mm sidewalls with variable widths from 1,500 mm to ≈2,500 mm. They are also equipped with a sophisticated computer-controlled system that gives accurate digital control of chemical output and automatic control of mechanical settings. This avoids operator error and ensures the accuracy of formulations programmed. Provision of extra chemical inlets into the mixing head with standard computer software allows for expansion at a later date without the necessity of mechanical changes or extensive electrical or software modifications.

8.10 CarDio process

The CarDio process is a method of manufacturing flexible PUR foam slab stock using liquid carbon dioxide as the physical blowing agent. This is additional carbon dioxide to that produced by the water/TDI reaction. This technology was developed by the Canon Group and installed on existing Maxfoam technology. The reasons for the use of liquid carbon dioxide are well known and this technology is used in many countries. The product is abundantly available in nature, is extremely cheap, expands three times more than competing alternatives, so less is required and has no harmful effect on the health of workers or factory safety.

By using carbon dioxide (which is a byproduct of other industrial processes), no additional gas is added to the atmosphere, which makes this method environmentally friendly. The CarDio equipment is available complete and incorporated with a new continuous slab stock foaming line or installed as an addition to a customer's existing foam block production plant as a retrofit package.

8.11 Viscoelastic continuous foaming

Although the conventional method of producing viscoelastic foam is the discontinuous process (in which large blocks are made one by one), developments in recent years have made it possible for some machinery suppliers to offer high-tech continuous systems. Alternatively, viscoelastic equipment packages are also available. They consist of all the necessary additional metering units and electrical controls for integration into the existing machinery for foam block production. However, because of the very heavy nature of the viscoelastic foam blocks, foam height will be reduced from the standard 45–48 inches to a maximum of ≤30 inches.

8.12 Laboratory-scale production

Laboratory-scale small foaming machines are available from some machinery manufacturers and these may suit large-volume producers to help develop and improve new materials for testing before using them on their prime continuous lines, which could result in large-volume waste. For small-volume producers, the laboratory 'box' test should suffice.

8.13 Inherent waste factors

In continuous flexible foaming lines, there are inherent factors with which a producer has to contend. Five of the major waste-generating areas are as follows:
- Provided the formulations are correct, faulty metering of one or more of the components because of the inaccuracy/failure in the metering pump will result in waste.
- Waste is generated at the start and end of a production run.

- The sides and bottom will produce thick and irregular skins.
- The top will be rounded because of a meniscus formed during foam rise.
- There is a possibility of fire because of the faulty formulation/inaccurate metering.

8.14 Moulded flexible polyurethane foams

Moulded flexible PUR foams can be used for making standard and contoured pillows, integral skin products, seats and many other consumer and industrial products.

This is a moulding process in which flexible and semi-flexible products are made. This section will discuss on a basic method of production using diphenylmethane diisocyanate (MDI) and polyol in a two-component system of raw materials. In two-component systems, the isocyanate is often called component A and the polyol component B. A producer of PUR foam has the choice of purchasing polyols with the desired additives already blended or purchasing pure polyol grades and blending at the production source.

8.14.1 Component A

MDI normally appears as a transparent, pale-yellow to a reddish-brown liquid depending on the specific grade used. The strength or concentration of the MDI is called the isocyanate content and is also known as the reactive radical or the group. Nitrogen, Carbon, Oxygen content (NCO) is the chemical shorthand way of saying isocyanate, just like H_2O means water. The isocyanate content measures the amount of reactive material in the MDI (component A). NCO content is important because it determines how much polyol is needed to complete the reaction. MDI and polyol must be combined in the correct proportions to prevent inferior parts being produced. If MDI becomes contaminated with water, the NCO content will drop because water or other foreign substances can react with MDI and weaken it chemically. If contaminated isocyanate is used and the machine settings are based on the original NCO content, then the imbalance between the two components will produce defective parts.

8.15 Effects of temperature on diphenylmethane diisocyanate

In general, the MDI used for PUR is a liquid at room temperature. Like all liquids, MDI has freezing and boiling points. MDI can thicken and solidify at temperatures below room temperature. Freezing can damage or ruin MDI, so correct storage temperatures as recommended by the suppliers should be maintained. Although MDI will 'boil' only at very high temperatures (usually 120–150 °C), some harmful vapours may be released as the temperature increases.

8.16 Polyol blends

A polyol blend or component B may have many additional ingredients such as chain extenders, fillers, catalysts, surfactants and blowing agents already mixed in. Just like the NCO content is used to measure the chemical strength of MDI, there is a similar number for the polyol. This number is the hydroxyl number or OH number, which is the chemical shorthand name for hydroxyl groups. These numbers will determine the exact quantities of MDI and polyol blend to be combined.

8.17 Basic chemical reactions

The basic reaction to make PUR is as follows:

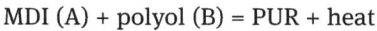

MDI (A) + polyol (B) = PUR + heat

MDI and polyol must be combined in exactly the right ratio otherwise the reaction will be incomplete and defective parts will result. Most of what goes wrong in production happens if changes occur in the chemicals, incorrect metering of components or in the faulty operation of equipment. MDI reacts very readily with water:

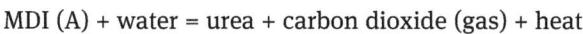

MDI (A) + water = urea + carbon dioxide (gas) + heat

Gas bubbles formed by the reaction of MDI and water cause the PUR system to foam. Only water should be allowed to react with MDI, which is the water already contained in the polyol blend.

8.18 Shipping containers for component systems

Different suppliers may have different types of containers but the most standard and common shipping containers for components are described below:
- *Steel drums* with closed or removable lids. With closed-lid drums, the lid is fastened permanently to the drum and the contents are removed through the larger bung hole in the lid. A smaller bung hole in the lid is also provided, and it is

advisable to open this slowly to let out any pressure build-up inside before removing the contents. In drums with removable lids, they can be taken off to remove the contents.

- *Totes*: There are several different types: steel, plastic in a steel cage and plastic in a cardboard box are common. Totes are also called bulk drums because they hold large volumes.
- *Bulk*: Some manufacturers who prefer to take price advantages of bulk purchases have very large chemical storage tanks that can hold >20,000 litres of each chemical. In this case, MDI and polyols are delivered in tank trucks and pumped through a closed line from the truck directly into the storage tanks. There are also other bulk tanks of lesser capacities (≈5,000 l) that are probably the norm for the production of moulded PUR foams.

It is advisable to check the delivery papers against the materials received, especially the certificate of analysis, which gives important information about the chemical make-up of each material, such as the NCO content for MDI and the OH number for the polyol resin. This information is valuable later on for calculating the correct mixing ratio. Spot checks, like opening one or two drums of the MDI to see the appearance, would be helpful. A flashlight might help to look into a drum. MDI should be clear, not cloudy. If it is cloudy, it may have frozen. Depending on the product grade, the colour can range from water-like to amber or brownish. If MDI is delivered in totes or tank trucks, a sample of ≈5 l can be taken for this check. If further checks are desired, a chemical check such as a hand mix can be carried with small samples of each component.

8.19 Checking for water contamination

If MDI appears cloudy or its surface has skinned over, it may have been contaminated with water. MDI and water react to form carbon dioxide gas which, in a confined space like a drum container, can build-up pressure and rupture the container violently. If MDI has been contaminated, it can be salvaged by transferring MDI to another container using a filter to catch solidified material. Disposal of this toxic material should be done according to standard safety procedures.

8.20 Unloading and storage of chemicals

After the preliminary checks have been completed, unloading the chemicals into the storage area is the next step. Even though it may seem like a routine job, only trained people and those under experienced supervision should handle this job. Improper handling can damage the chemicals or injure workers. If a spill occurs, emergency procedures must be followed by trained personnel. During the unloading of bulk chemicals, each chemical must go into the correct storage tank: MDI to the MDI tank and the polyol resin to the polyol tank. It would be dreadful to think what would happen if one or the other went into the wrong chemical tank. Although it is very unlikely, it is mentioned here because there have been cases where this has happened. If this unlikely event takes place, all personnel in the factory should be evacuated and outside emergency professional assistance must be called for.

Storage of chemicals should be according to standard procedures to enhance safety. Protection from extreme temperatures and possible contamination of water vapour should be provided. Keep all drums and totes tightly sealed. If any were opened to do the receiving checks, re-seal them for storage. Never re-seal a drum or tote that has been contaminated with water. The carbon dioxide gas produced can create tremendous pressure. Store both chemicals at temperatures between 75 °F and 90 °F. Keep the MDI drums and totes on pallets because cold concrete floors can cause freezing, even if the room temperature is not cold.

Treat materials in bulk tanks with extreme care because they hold very large amounts of chemicals and the results of improper storage can be very expensive, very dangerous or both:

- *Gas blanket*: In general, MDI tanks have a head space (empty space between the top of the liquid level and inside of the tank) which is filled with dry air or an inert gas, such as nitrogen. The head space must be kept filled with this gas blanket to prevent the water vapour in the air reacting with MDI.
- *Temperature control*: Bulk storage tanks are usually connected directly to the PUR processing equipment. Because the right temperature is important in processing, the material in the bulk storage tanks must be kept at a controlled temperature.
- *Recirculation* is important for temperature control. By pumping the material at a slow rate from the tank, through the pipelines and back to the tank, a constant temperature is held through the whole system. If the material just sat in the tank, the portions in the distant parts of the piping system would quickly get too cold. This could cause processing problems at the machines or result in a blockage caused by frozen material.
- *Mixing*: Unlike the material in the very big storage tanks that does not need mixing, the polyol resin material in the drums, totes or smaller bulk storage tanks must be mixed thoroughly before use to ensure a homogenous mixture of the polyol blend by preventing 'settling' of additives at the bottom. A slow rate of mixing and possible recirculation will be adequate and, in some instances, outside temperature-control devices are used.

8.21 Preparing for production

Before the two components can be used for production, the polyol should be mixed thoroughly to ensure that all ingredients in the polyol will be a homogenous mix. For this operation, three types of mixers are commonly used:
- *Propeller mixers* probably give the best mixing. These are used with open-lid drums. Mix the contents in the drum for ≥30 min before use.
- *Bung-hole mixers* have smaller blades that can fit thorough a bung-hole of a closed-lid drum. Because the blades are smaller, bung-hole mixers do not work as well as propeller mixers, and mixing should be for a longer period, say 45 min.
- *Emulsifying high-shear mixers* work at high speed and, even though they shorten the required mixing time, they should not be used because of the rapid heat build-up that can harm the material.

Handling polyol resins in totes is very similar to handling resins in drums. Mixing is slightly different with totes. It is best to use only propeller mixers. Bung-hole mixers are too small for the large totes and emulsifying mixers can damage the material. Because of the large volume of material, it is best to mix it for ≥1 h at slow speed and continue doing so while the tote is in use. The addition of colourants can be achieved in two ways. Some manufacturers may want to inject the prepared colorant directly into the dispensing mixhead. Most manufacturers prefer to add the colourant to the polyol resin before it is added to the day tanks. The correct type of procedure for adding colourant will depend on the type of container (drum, tote or bulk) and the type of colourant. Ensure that the colourant is mixed thoroughly in the resin to prevent product rejects because of colour distortion. If additives are added at source or additional additives are added after receiving the material, it should be done before the original mixing starts. The correct amounts as specified by the formulation must be added so as not to alter the final properties of the foam.

8.21.1 Filling the day tanks

Using two-component raw-material systems is that much easier because only two large day tanks have to be prepared. Smaller operations may even be able to work directly from the drums but, for good control of the material, the day tank system would be best. If the same grades of materials are used, filling the day tanks will be easy. However, if different grades are to be used, then each day tank must be thoroughly flushed at least twice to remove the old material. Once the day tanks have been flushed properly, they can be filled. When filling the MDI tank, special care is needed to prevent water contamination, especially from the air. After filling this tank, make sure it is completely sealed and pressurised with dry air or nitrogen.

8.21.2 Calibration of processing equipment

Processing equipment is available from many suppliers with standard or highly advanced features, but the general operating principles are more or less identical. Calibration of the processing equipment before production is a very important aspect of the operation to ensure good-quality parts. Different two-component systems will have different mixing ratios and will need different machine ratio settings. Before setting the machine ratio, make sure the materials are at the correct processing temperatures. If the temperature changes during or after calibration, then the machine ration will have to be reset.

Each type of machine/equipment will have its own recommended method of calibration. Some machines are calibrated directly at the mixhead, whereas others may have separate valves for calibration. The operational manuals will provide instructions for the appropriate calibration procedure. To calibrate, an average production shot is poured into pre-weighed cups. The cups are then weighed again to find out how much polyol and MDI are coming out of the machine with each shot. An average of the amounts of polyol and MDI are taken and the machine ratio determined.

If the machine ratio is different from the predetermined mixing ratio, the output of the pumps needs to be adjusted until the machine ratio matches the required mixing ratio. The simplest way is to increase or decrease the pump speed of the polyol, MDI or both. If the pumps are gear-driven, increasing the number of teeth on the drive gear will increase the pump output. Increasing the number of teeth of the pump (idler) gear will decrease the pump output. After the machine has been calibrated, the last step before running production is the cup shot or free rise foam check. The cup shot test is done to make sure that the foam reactivities are normal. The cup shot test can catch many of the common chemical and equipment problems that can lead to production rejects. A cups shot is done by pouring a small shot from the machine mixhead into a pre-weighed paper cup. Pour in just enough material so that the foam expands slightly above the top of the cup when it reacts. Four tests can be done with the cup shot – foam reactivities, throughput, free rise density and foam structure.

8.21.3 Foam reactivities

The following describes the different stages of the foaming process:
- *Cream time* is the time in seconds, from beginning the pour until the material starts to expand. The liquid mix starts to change to a milky colour and starts to rise.
- *Rise time* is the time from beginning the pour until the foam stops rising. By this time it should look like a PUR foam 'muffin'.

- *Tack-free time* is the time from beginning the pour until the foam surface is no longer tacky. Use a clean wooden stick to touch the surface lightly and, if it is still tacky, the tack-free time has not been reached.
- *Pull time* is the time from beginning the pour until the top of the foam can be pinched and pulled without tearing.

These reactivity times are like a 'fingerprint' for two-component raw material production systems. Different two-component systems will have different standard reactivity times and, if the chemistry is correct and the machine is working well, the same results should occur for different number of pours. If the times measured are different from the standard, then it is an indication of chemical or machine problems. In general, suppliers of raw materials give an indication of these times that may be used as guidelines or standard. Different dispensing machines can have different reactivity times with the same two-component system, so calibration and testing becomes important.

8.21.4 Throughput

Throughput is measured by weighing the cup before and after foaming and then dividing by the length of time the material flows from the mixhead. This is known as the shot time and is measured in grams per second.

8.21.5 Free rise density

Free rise density measures the foaming power of the particular two-component system used and gives an idea how it will fill out the moulds in production.
      After doing the reactivity tests:
- Cut off the top of the reacted foam so that it is flush with the top of the cup.
- Weigh the cup and subtract the weight of the empty cup to get the weight of the foam.
- Divide the weight of the foam by the volume of the cup to get the free rise density.

Calculate the volume of the cup by weighing the amount of water needed to completely fill it. The weight in grams equals the number of cm^3 in the cup. Then, the free rise density can be expressed in g/cm^3.

8.21.6 Foam structure

Check the foam structure by cutting the cup lengthwise in quarters. The foam should be uniform and have a smooth, velvety texture. Splits, swirls or other imperfections are usually signs of machine or equipment problems. For solutions to these problems, refer to the troubleshooting guide shown later in this chapter. Once these tests are done and the results are satisfactory, then production can begin.

8.21.7 Filling the moulds

The heart of the production process is the mixhead, which combines the polyol and MDI and pours or dispenses the mixture into a mould. There are two main types of equipment set ups for filling moulds – open pour and direct-injection:
- *Open pour*: The mixed PUR system is poured into an open mould. The mixhead is moved clear of the mould and the mould closed. Open-pour moulding can be done by hand or the mixhead can be machine-controlled.
- *Direct-injection*: In this case, the mould is closed and the material injected into the mould through a small opening/hole or an injection port.

If machine-controlled open pouring or direct-injection filling of moulds is used, the operation is easy and simple. If however, open pouring with hand pouring is used, more control of how the moulds are filled is needed.

Hand pouring: The way a mould is filled is very important. Even if everything else is perfect, the wrong pouring pattern can cause rejects. Some of the common problems that can result from a bad pouring pattern are as follows:
- Flow lines
- Air bubbles because of trapped air
- Mould not filled
- Hard and soft areas in same part
- Knit lines
- Mould not venting properly

8.21.8 Pouring pattern tips

The main purpose of a good pouring pattern is to make sure the mould is filled evenly and smoothly. Finding the right pouring pattern is sometimes tricky and will take some skill. The same pattern may not necessarily work for different part designs. Trial and error may be the only way to find out, but there are a few guidelines that will help to find a good pouring pattern faster:
- *Never back-track*: Always pour in one direction only. Back-tracking will lead to knit lines or other problems in the finished part.
- *Follow the mould design*: If the part has some areas that are thicker, these will need more material, so pour a little extra there. Likewise, thin areas should get less material.

- *Watch for tilted moulds*: If the moulds are tilted, the material will run downhill before it can react and fill up the mould. In this case, pour a little extra material in the uphill part area of the mould to make up for what will flow downhill.
- *Pay attention to vents*: Most moulds are 'feather-vented', that is, there is a slight clearance between the two mould halves running the whole way around the mould. When the liquid mix begins to foam up, the air in the mould is pushed out through this small gap. If a mould is not-feather vented, it will have notched vents to let the air escape. Because the foam can also come out from these vent holes, the pouring pattern has to be adjusted for these types of moulds. Usually, the practice is to start pouring from the opposite end to where the vents are.

8.21.9 About moulds

Some moulding systems can control temperatures. This is important for the chemicals as well as for the moulds. If the mould is too hot, the finished product will have blisters on the surface and the part will be a reject. If the mould is too cold, the finished part may be difficult to get out of the mould or it may tear when de-moulding.

8.21.10 Mould-release agents

Mould releases are an essential part in flexible PUR foam mouldings. In addition to easy removal of parts, they assist and enhance product quality in many ways by helping in material flow in the mould and also improve the surface quality of the finished parts. Although mould-releasing agents can have an effect on the quality of a part, they do not affect the chemical mix or the mixing ratio, and are just like 'spectators' to the chemical reaction taking place. Mould releases are sprayed into the mould by hand or automatically with a device. Mould-release products differ from manufacturer to manufacturer, so the supplier's instructions should be followed.

In general, only a very light coating to cover all the insides of a mould is required, and it may not be necessary to spray with every moulding cycle. Some moulding cycles, depending on the product, may need mould release only every 3–4 mouldings or even occasionally. Too much mould release can be as bad as too little. Using too much mould release can cause surface quality problems, and using too much is a more common mistake than using too little. There are two ways to tell if too much mould release is used: obvious puddles in the mould after spraying or surface quality problems when parts are de-moulded. Problems such as wet spots, pinholes, streaks and a hazy appearance can be signs that too much mould release was used.

8.21.11 In-mould coatings

An in-mould coating is a special type of paint. Instead of painting a part after it is moulded, the coating is applied to the inside of the mould before the chemicals are dispensed from the mixhead. After the foam reaction is completed, the part comes out of the mould already painted. This sounds easy enough but it is advisable to initially seek advice and guidance from suppliers and technical experts. Some of the in-mould coatings have mould release built in so they do not need a separate mould-release spray. If not, a separate mould release is sprayed into the mould, then the paint is sprayed and then the foam moulded. The principles of spraying in-mould coatings are the same as for applying mould releases.

8.21.12 Moulding with inserts

Inserts are anything that is moulded into the PUR part. Inserts can be used for many reasons. They can be used to give more support to the soft foam or to act as an attachment point for screws and bolts. Plastic plates and metal rods often serve this purpose in furniture parts. Inserts can also add special features to the foam (airbags and gel bags moulded into soft footwear are good examples) or they can be used just to fill up space.

There are three basic things to keep in mind when working with inserts – temperature, contamination and positioning:

- *Temperature*: Inserts should be heated to the same temperature as the mould before being placed into the mould. Avoid injury by wearing gloves. If the inserts are too cold, the finished part will have sink holes and will probably be a reject.
- *Contamination*: Inserts must be completely clean. Grease, oil or mould release can cause a part to be rejected. Foreign substances can also destroy the bonding between the insert and the PUR foam.
- *Positioning*: Inserts must be positioned correctly in the mould. If pins or clips are used to hold the insert in pace, make sure the insert is held firmly. Inserts that are not securely in position may be bumped out of place by the force of the reacting foam inside the mould. This will lead to rejects.

8.21.13 Mould clamping

When PUR foam systems react in a mould, they create high pressure as they expand. Moulds must be clamped shut with a pressure that is higher than this internal mould pressure or the expanding foam will force open the mould. If this happens, the foam will flow out of the mould and create rejects. Moulds can be clamped mechanically or hydraulically. When moulding PUR foams, make sure the clamping systems are working correctly and the mould is completely shut after each shot.

8.21.14 Mould cleaning

Over time, moulds become dirty from the build-up of mould release and foam residue. Even before it is visible, this thin film can affect the surface quality of the moulded parts and possibly lead to rejects. This is especially true with highly detailed parts. The first sign of a dirty mould is when the parts are de-moulded. A whitish build-up may be seen on the parts. Also, dirty moulds might make de-moulding difficult by causing parts to stick in the mould. In extreme cases, a whitish build-up will be seen on the moulds themselves. Mould-cleaning chemicals may be aggressive chemicals, in which case adequate precautions must be taken.

8.21.15 Flushing the mixhead

Use flushing agents to keep the mixhead clean. During production, reacted material can build up on the mixhead and eventually cause blockages or other mixing problems. If the machine is stopped often, the build-up can happen very quickly. Most machines have a specific control button for flushing or purging the mixhead. Most machines flush by forcing flushing agents and high-pressure air through the mixhead. Usually the mixhead is flushed into an empty drum, and used specifically to collect the flushing agent. Make sure that this drum is well away from the moulds to prevent splashing or spraying this flushing agent into the moulds. After flushing, make sure the mixhead is wiped clean (if necessary) to prevent the flushing agent dripping into the moulds when the pouring operation starts again.

8.21.16 Lead-lag

Lead-lag occurs if the polyol and the isocyanate do not enter the mixhead simultaneously. One component leads and the other lags behind, resulting in a poorly mixed area of foam. This condition usually appears at the start of the pour and can result in various problems. To do a lead-lag test, carry out the following:
- Pour a thin layer of foam onto a polythene sheet, making sure to separate the start and the end of each shot.
- Allow the foam to cure for a few minutes.
- Peel the foam 'pancakes' from the polythene sheet.
- Brush the bottom side of the foam (side that was in contact with the sheet) with a 20% solution of hydrogen peroxide. Hydrogen peroxide acts as an indicator for the component that is leading or lagging. It is similar to how a pH test determines the acid/base level in a pool. After brushing with the hydrogen peroxide, one of the following can be seen:

- Entire surface turns pale yellow. Indicates there is no lead-lag.
- A dark-yellow stain appears around the initial shot point. Indicates MDI lead.
- A white halo appears around the initial shot lead. Indicates a polyol lead.
- The surface is covered with alternating white and yellow streaks. Indicates poor mixing.

Note: Check the troubleshooting guide for recommended solutions.

8.21.17 Recharging day tanks

After running the machine for several hours, the levels of the polyol and isocyanate in the day tanks should be checked for refilling or recharging long before they are empty. Once the levels have gone down below one-quarter of the capacity, holding a steady temperature is difficult for most machines. Changes in material temperatures can lead to processing problems, so tanks should always be above at least one-quarter the capacity level. As long as the same material grades are being used, flushing the tanks before recharging is not required.

If it is necessary to switch to another two-component system, it is important to follow the standard changeover procedure. The new system will have a different chemical composition, which means a new mixing ratio has to be calculated and set on the machine. Also, the viscosities of the two components may be different and the machine pressure has to be adjusted accordingly. If all these adjustments are made correctly, the new system should run smoothly and prevent unnecessary downtime. The most important steps to remember are as follows:
1. Flush out the old system with the new material.
2. Set the right temperatures.
3. Calibrate the equipment and set the correct mixing ratio.

Note: For safety factors and general troubleshooting recommendations, refer to the relevant chapters.

8.22 Examples of typical processing

Exercise A: To make a conventional flexible foam block of size 1.85 × 1.1 × 1.0 m having a density of ≈28.8 kg/m^3 by the discontinuous method and manual pouring.

Process: Prepare a large wooden mould with ≈1.3 cm all around for shrinkage and an allowance for extra height. This mould can be made of other materials such as thick aluminium, thick plastic sheet or other suitable material, and the four sides should be detachable for easy de-moulding. If the mould is provided with wheels it helps movement. The mould sides should be waxed or sprayed with a release agent or lined with thin polyethylene film and made leak-proof. There are two simple ways to mix the components.

Some may opt to place a large mixing vessel on the bottom centre of the mould with a lifting device (electrical or chain block) or have the mixing vessel outside and pour manually. The former is the better method but one must ensure that the mixture will not leak from the bottom during mixing.

Consider Table 8.1 as a guideline:

Table 8.1: Foam formula.

Component	Quantity
Polyether polyol	31.20 kg
TDI	21.35 kg
Water	1.83 kg
Methylene chloride	2.21 kg
OS20	0.56 kg
PS 207	0.06 kg
SO	0.06 kg
Colour	0.03 kg

Weigh accurately all the components separately. Pour the polyol into the mixing vessel and put in the colour and mix for 40–60 s with a high-speed mixer. Mix the OS20, PS207 and water separately and add to the polyol and mix for 15 s. Now add the methylene chloride and SO and mix for a further 15 s. Finally, add the TDI and mix for 3–4 s and pour or lift the mixing cylinder off the floor of the mould. The mixture must be poured onto the floor of the mould in the liquid state before it has started to 'cream'. The liquid mixture will spread slowly and begin to rise. If a flat top is desired, place a very light floating lid on top of the rising foam when it has reached about one-third of the mould height. This will prevent the rounded meniscus being formed on top. Allow ≈10 min for the foam to settle and achieve its first phase of cure and, when the foam top is tack-free, the foam block can be de-moulded and transported to the final curing area. Follow standard storage procedures and after 24 h the block will be ready for fabrication.

Note: Although this method may be considered 'out-of-date' because high-tech machines are available, this practice goes on in many developing countries where the market volumes are not very large. It is an effective way of making good foam for a small producer with limited starting resources.

Exercise B: To make high-resilience cushions for office and domestic furniture or similar products by manual hand-drilled or a flexible PUR dispensing machine, the density range will be ≈55–60 kg/m^3 and is based on using a typical two-component raw material system whereby adjustment of the ratio enables different densities to be achieved.

Typical data: Hand or machine mixing speed = 3,000 rpm.
Mixing ratio = 100:55
Mixing time = 6–8 s
Cream time = 12 s
Gel time = 60–80 s
Tack-free time = 5–8 min

8.22.1 Material storage

In general, two-component systems are supplied in steel drums with two bung-holes. It is customary that drums containing each component are coloured in two colours (e.g. blue/red or green/red) for easy identification and storage. Standard storage practices should be followed. Here, the storage life of each material must be considered, which will be provided by the supplier.

8.22.2 Handling of raw material

When handling the raw materials, it is essential that protective wear is worn always and that the factory has a spill-management procedure. The isocyanate drums should be stored in a dry environment indoors in a well-ventilated area and stored at 20–30 ° C. When opening a container of polyol or isocyanate, care must be taken to release any internal pressure slowly by opening the smaller bung-hole slowly. The larger bung-hole can be used for pumping out the material but, before doing so, the polyol must be mixed thoroughly to ensure a homogenous blend, otherwise bad parts may result. Particularly in the case of isocyanate, it is good practice to introduce some nitrogen into a drum after the material has been taken out because the balance material in the drum will react with the air.

8.22.3 Moulds

Moulds can be fabricated from various materials such as metals, plastics, timber or fibreglass. Remember that there is a pressure build-up inside a mould during foam rise and the mould must not warp or give way. To counteract this, moulds should be provided with a few strategically placed small vents ($\approx$1 mm). This allows any contained air and gas to escape. If vent holes are too large it may affect the required pressure within the mould and result in a bad foam (e.g. foam shrinkage or imperfections on moulded foam surfaces). It is ideal to maintain the mould temperature at 35–40 °C, which will aid in curing and give the moulded foam parts a smooth silky finish.

8.22.4 Mixing and weighing procedures

It is always a good practice to carry out a small 'box test' in the laboratory before production. To produce good-quality foam, each component should be weighed accurately in clean and dry containers. If colour is desired, it can be mixed into the polyol component. As mentioned above, the polyol blend must be mixed mechanically before being taken from each drum. If an electric drill is used for mixing, make sure it can mix at ≥3,000 rpm. Slower speeds will produce poor-quality foams. The mixing should be done quickly and the mixture (still in the liquid state) should be poured or dispensed into the mould before 'creaming' takes place. If the ambient temperature at the time of production is too high, the creaming time is faster and, in general, a larger batch size will react faster than a smaller batch.

8.22.5 Tips for minimising waste

In the production of flexible PUR, irrespective of the methodology, machinery and equipment used, there is a certain inherent waste factor, particularly for slab stock foam. Whether the foam blocks are in single or continuously foamed blocks, the hard skins (bottom, top and sides) are considered waste and this is in addition to the process waste that could be generated because of various reasons (particularly faulty formulation or weighing/metering). Foam waste as a standard may be acceptable at ≈10–15% but even some large-volume operations may be operating at waste factors >30%, which is unacceptable. True, the waste can be recycled into some products and in this way a producer will get a kickback but the possibilities are limited:

- In practice it has been found that cooler temperatures of ≈20–25 °C give better foam than at 28–30 °C. That is, in tropical countries, it is better to 'foam' in the mornings and evenings than during midday unless the temperature can be controlled. In colder countries, it is best to have the atmospheric temperature controlled when foaming.
- Use a light 'floating lid' to flatten the top in the manufacture of single foam blocks. In continuous foaming, a flattening device will help to flatten the top. Most sophisticated foaming lines can now produce continuous 'flat tops'.
- If these devices are not available and the foamed blocks have rounded top surfaces, soon after foaming, stack the blocks upside down before the final curing stage. This will reduce the top waste.

- In continuous foaming, there is considerable waste at the beginning and end. These are called 'end' blocks. Instead of the customary short runs, longer foaming runs will reduce the waste considerably. Also, without rejecting the end blocks, if they are converted to small cushions and sponges, there will be a kickback and waste reduction.
- It is the normal practice of continuous foaming to cut into standard size 'buns'. This practice will generate waste. All foam manufactures know in advance of orders in hand for sizes needed for mattresses, cushions and slabs, so staff in the fabrication division can work out the exact sizes of blocks needed and request the production line to cut the buns into these sizes, which will be different in length. In this way, the buns can be converted to products with little or no waste.
- Make sure the moulds are set up appropriately. In the case of single-block production, if polyethylene film is used as a mould liner, make sure there are no excessive creases that will 'eat' into the foam sides and create added waste. All weighing/metering and mixing times must be accurate to prevent reject blocks. In the case of continuous foaming, make sure the metering pumps and the mixing/dispensing head are in good working order so as not to produce bad foam or the possibility of foam catching fire in the trough. The paper making up the trough should be set up appropriately so as not to tear during the foaming process and make the foam leak, thereby creating waste.

9 Basic safety factors

Safety starts with the building layout for a polyurethane (PUR) foam factory. Because of the highly flammable nature of the raw materials and the corrosive and harm caused to the personnel, safety is of prime importance. This is stated not to create alarm but to stress the importance of adhering to basic rules of safety to ensure smooth and uneventful production with maximum protection for operating personnel. Material safety datasheets (MSDS) are an internationally accepted safety standard that provide very useful information to follow.

9.1 Buildings

It is ideal, as practiced by large-volume producers, to have separate buildings for housing the raw materials, storage tanks, production area, post-cure holding area, cutting and fabrication area and the finished products/shipping area. These should be well planned with concrete floors, sufficient emergency exists in case of danger and also with easy access to each other to minimise transport activity. Naturally, a good layout plan will have a 'flow-pattern' for efficient productivity. Small-to-medium producers of foams may not have the luxury of separate buildings but, because their operations are small, they can manage to cope but must at least separate the raw materials and post-cure holding area from production (even in the same building). In both cases, the provision of efficient fire extinguishers at strategic points is crucial. No smoking is allowed in any area at any time (including the recreation room if it is within the building).

9.2 Storage of raw materials

There are many safety methods for storing and handling of raw materials. Good ventilation, anti-static conditions, temperature control and trained personnel for handling the materials are some of the basics. Raw materials should not be left outdoors at any time, and most suppliers provide information on safety factors as well as the shelf-life for each material. If the raw materials come in drums or totes, they will have colour codes for easy identification.

9.3 Production

All workers must be protected from the most common exposures to chemicals. Special care should be taken if handling isocyanates and polyols; if contact is

https://doi.org/10.1515/9783110643183-009

made, immediate attention (e.g. washing with water, a shower, first aid or even medical help depending on the severity of the problem) is required. All personnel working with these chemicals must be made aware of the possible dangers. Supervisors must be trained in handling chemicals safely and also the correct actions to take in case of an incident or more serious accident:

- Vapour inhalation: wear breathable masks or respirators.
- Eye contact: wear goggles or safety glasses.
- Skin contact: wear long-sleeved gloves/protective wear.
- Vapour indigestion: never bring food or drink into production, storage or laboratory areas.
- Spills: report and clean up immediately.
- Possible fire: if small, use fire extinguishers, otherwise call for outside assistance.

9.4 Safety provisions

PUR chemicals also pose health hazards so should be handled very carefully. Safety factors start from the storage point of raw materials to the full cycle, that is, until the finished products are sent to the holding store for the finished goods:

- Provide eyewash stations and showers.
- Provide adequate exhaust ventilation where needed.
- Keep MSDS at strategic locations.
- Provide thorough training in safety factors/procedures.
- Fire hazards: local fire officials will carry out periodic training on request.

9.5 Basics of spill management

Chemical spills may not be a frequent occurrence in a well-managed factory, but they may happen. The following recommendations help if this occurs:

- Identify the material.
- Clear the area of non-emergency personnel.
- Notify the supervisor.
- Wear air respirator/mask/safety glasses, gloves, protective suit, boots.
- Stop the spill/leak.
- Contain the spilled material.
- Absorb the spilled material with absorbent or sawdust.
- Use a solution from a spill cleanup kit to neutralise the spilled material.
- After ≈15 min, shovel the material into a container.
- Let the container stand until neutralisation is complete.

– Dispose of the neutralised residue in accordance with local laws.

Note: For large spills, external professional assistance may be needed.

9.6 Recommended safety equipment

All operations dealing with chemicals need safety equipment and safety wear. There are many different types, and the following are some of the recommended ones:
– Protective gloves
– Protective wear
– Safety glasses and goggles
– Respirators
– Safety shoes
– Eyewash stations
– Showers
– First aid kits
– Fire extinguishers
– Spill control kits
– Safety signs
– Air monitors

Note: There are many types of the items described above, and a foam producer must select the type and sizes as best suited for his/her operation. Some may opt to forego some of the items described above and include different items. The main idea is to have sufficient and essential safety equipment to ensure a smooth, efficient and trouble-free manufacturing operation, giving due importance to safety factors at all times.

PUR chemicals are termed 'hazardous' and it is important to follow established basic guidelines in the appropriate handling of hazardous materials, procedures for disposal of the chemical wastes generated and methods for the containment and cleanup of spills. This necessitates sound knowledge of the safety factors involved, and all supervisory staff and management involved in the different operations of handling and using these chemicals should be well versed in all aspects of safety.

9.7 Handling precautions

To minimise the hazards associated with the handling, use and storage of PUR chemicals, the following are recommended:
– Always wear appropriate protective equipment.
– Never work alone if handling or using hazardous/reactive chemicals.
– Do not inhale vapours, mists or dusts.
– Avoid contact of all formulation chemicals with the skin and eyes.

- Handle freshly polymerised foam with care. Be aware of vapours during reaction/cure.
- Do not stack fresh 'buns'. Excessive congested heat can cause spontaneous combustion.
- Equip foam storage areas with efficient sprinkler systems.
- Keep adequate quantities of isocyanate neutraliser to counter spills or leaks.
- Never expose isocyanate held in containers to water, amines or other reactive chemicals.
- Never expose PUR chemicals in closed containers to elevated temperatures.
- Never expose PUR foam to an open flame or high heat source.

9.8 Health and industrial hygiene

9.8.1 Polyols

Typically, polyols are mild irritants to the eyes and skin and exhibit very low acute oral toxicity. Despite their relatively innocuous characteristics, they should be handled with appropriate caution, and contact with the eyes or skin should be avoided at all times. Eye protection and elbow-length gloves are recommended if handling polyols. Skin in contact with polyols should be washed immediately with soap and water; any contact with the eyes necessitates flushing the eyes with large amounts of flowing water. If other chemicals are blended into a polyol, the ventilation and handling requirements should at least conform to the requirements for the most hazardous chemical in the blend.

9.8.2 Isocyanates and prepolymers

Isocyanates and prepolymers can be hazardous unless recommended safety precautions are understood and practiced diligently. MSDS will provide this information, but so will chemical suppliers. The goggles of chemical workers should be worn if they are handling isocyantes. All isocyanate work areas should be equipped with an eyewash station. If an isocyanate comes into contact with the eyes, first aid must be administered immediately with the eyes held open while flushing with a low-pressure stream of water for ≥10–15 min. Medical assistance may also be sought as a safety precaution. Repeated or prolonged contact with the skin may cause irritation, blistering, dermatitis or skin sensitisation. Inhalation of vapours may cause respiratory problems. In the event of direct contact with skin, use a shower immediately. Wash the affected areas of skin thoroughly, using plenty of water and a mild soap. Contaminated clothing should not be worn again until laundered.

Isocyanates are not likely to be swallowed in standard industrial operations and exhibit low oral toxicity, but some materials may burn the tissues of the mouth, throat and stomach. If isocyanates are ingested, medical attention should be sought immediately. Isocyanate vapours and mists are irritating to the nose, throat and lungs. Sometimes, even brief exposure can cause irritation, difficulty in breathing and coughing. Toluene diisocyanates (TDI) have a relatively high vapour pressure at ambient temperatures and a vapour density six times that of air. Thus, open containers that release TDI to the atmosphere result in a relatively high concentration of TDI vapour that does not disperse readily. Sensitisation to isocyanates may result from excessive exposure, and subsequent exposure to very low concentrations might provoke an allergic reaction with symptoms such as asthma. Methylenediphenyl diisocyanate (MDI)-based isocyanates have a lower vapour pressure and do not become airborne readily at room temperature. However, mists are formed easily and should be treated with the same concerns as those discussed previously.

9.8.3 Component systems

The polyol side of a system is typically not as hazardous as the isocyanate side. However, caution is recommended. Formulated polyol systems often contain volatile blowing agents or catalysts such as amines, which may pose hazards under certain conditions. Good ventilation and appropriate handling practices as recommended by MSDS for these chemicals should be followed. If systems contain low-boiling fluorocarbons, they must not be stored at temperatures that may cause pressure build-up and result in container rupture.

The isocyanate side of a system is hazardous. Different systems may use different isocyanates. The most common isocyanates are based on TDI or MDI, which may be present as prepolymers, polymeric or crude materials, and should be treated with equal importance. Inhalation is the most common cause for exposure, so it is advisable to be aware of the vapour pressure of the products. All isocyantes will have a vapour hazard in direct proportion to the amount of free monomer (TDI or MDI) present. Use at elevated temperatures will increase the vapour hazard.

9.8.4 Other raw materials

Additives such as surfactants, catalysts, flame retardants, fillers, blowing agents and pigments are used in PUR foams. Each has its own profile of hazards and recommended safety precautions, and suppliers will provide this information and also specific data on safe handling, use, storage and disposal.

Most polyols have low volatility and flash points ≈125–220 °C. They are classified as class-B combustible liquids. Being organic materials, they burn under fire conditions to liberate water and carbon dioxide and, if sufficient oxygen is present, some carbon (soot) and carbon monoxide. Polyols have no known explosion potential but, as with all organic substances, if heated to decomposition in a confined space, they may generate sufficient volatiles to be a potential explosion hazard.

Most commercial isocyanates have a high flash point and are classified as class-B combustible liquids. They may burn in the presence of an existing fire or high heat source and adequate oxygen. The low volatility of isocyanates minimises the potential hazards of explosion but, under fire conditions in which a large concentration of isocyanate vapour is generated, an explosion can occur. Carbon dioxide or dry chemical extinguishers are effective for fighting these fires. Water spray in large quantities is also effective for large areas. When fighting these fires, protection is needed against nitrogen dioxide vapours and isocyanate fumes. The most common deterioration in the quality of isocyanates is exposure to water and heat.

If they come into contact with water, isocyanates react readily, producing heat, carbon dioxide and insoluble ureas. The pressure created by the evolved heat and carbon dioxide may be sufficient to rupture a closed container. To protect isocyanates from contamination from atmospheric moisture, all containers and equipment to be used should be cleaned carefully and purged with a dry inert gas such as nitrogen that is also preferably free of oil and dust. Isocyanates should also be stored in the recommended temperature range to avoid freezing. If freezing occurs, following the manufacturer's directions can restore the isocyanates to a usable state. At temperatures >80 °C, isocyanates may trimerise in an exothermic reaction to form isocyanurates, carbodimiides and carbon dioxide, which create pressure hazards inside closed containers.

Many liquid amine catalysts have flash points and are classified, according to their flash points, as flammable or combustible liquids. Because of their possible volatility or flammability, therefore, amine catalysts should be considered as fire hazards.

The commonly used auxiliary blowing agent methylene chloride has neither a flash point nor a fire point by generally accepted standards. However, if heated to decomposition level, it produces various potentially hazardous materials such as hydrochloric acid, chlorine, carbon dioxide and carbon monoxide.

9.9 Spill management

Isocyanates – Spills and leaks should be contained and cleaned up by appropriately trained and equipped personnel. In case of spills or leaks, all personnel in the area should be evacuated immediately. Standard protective equipment should include self-contained breathing apparatus as well as impervious clothing, footwear and gloves. An approved respiratory protective device must be worn if there is a possibility of exposure to isocyanate vapours exceeding the permissible exposure limit of 0.02 ppm or 0.05 ppm time-weighted average threshold limit set by the Occupational Safety & Health Administration.

All leaks and spills should be contained immediately to prevent contamination of the surrounding areas. If the source of the leak is a damaged or leaking drum, remove the drum or drums immediately to a well-ventilated isolated area or outdoors. Carefully transfer the contents to a clean leak-free drum or container, and dispose of the damaged drum. If the source of the leak is a damaged storage or holding tank, plug the container with a rubber or wooden plug and transfer the contents into a clean and dry leak-free tank. Make sure the damaged tank is cleaned thoroughly before permanent repairs are attempted.

In cleanup procedures, sawdust or other absorbent materials and a liquid neutraliser are used. There are standard neutralising liquids, but a suitable neutraliser consists of 5% aqueous ammonia (sodium carbonate) and 1–2% detergent in water. Any standing pools of isocyanate should be pumped into closed-top but sealed containers for disposal. Cover the remaining isocyanate with absorbent and then physically remove it to a suitable container for further decontamination and disposal. After the spill area has been again treated with a neutralising liquid, this liquid should be removed with more absorbent material. Final decontamination may be achieved by spraying the area with large quantities of water. If possible, prevent isocyanate from entering drains because the reaction with water may cause blockages. Checks should be made to ensure that no residual active isocyanate adheres to the ground surface and, after air samples have been taken, the area can be declared safe and reoccupied.

Polyols – Small spills on hard surfaces can be absorbed with sawdust or other absorbents and then swept up for disposal. Large spills should be pumped into drums or containers for disposal. Personnel engaged in this operation should wear appropriate skin and eye protection.

One risk that arises from the use of isocyanates is inhalation of vapours. During any activity with isocyanates, it is essential to maintain adequate mechanical ventilation of the work area. If vapour concentrations cannot be controlled under recommended guidelines through ventilation, then respiratory protection is a must at all times.

Only thoroughly trained and appropriately equipped personnel should be allowed to participate in disposal operations. Waste isocyanates and waste polyols should be handled separately, and local and district regulations must be followed. There are also general regulations for the disposal of empty drums and containers used for PUR chemicals. General liquid wastes primarily comprise isocyanates, polyols and auxiliary chemicals associated with them in the foam-making processes. These may occur, for example, in the form of residues in empty drums, containers and unstable materials. Solid waste PURs are usually disposed of by incineration or landfill, and they may be useful for adhesive manufacturers. Other methods of waste disposal are reuse, re-bonding and recycling and, in this case, only the really unusable solid wastes will be the subjects for final disposal.

10 Setting up a manufacturing plant for an entrepreneur

Large blocks of quality flexible polyurethane (PUR) foams can be made success-fully using manual methods (Figure 10.1). For an operation of this nature, a suitable foaming system can be fabricated by an entrepreneur with basic engi-neering skills.

On the assumption that an entrepreneur will have limited resources and that small-to-medium volumes of foam manufactures are planned, several basic guides can be considered to be viable. These presentations are based on actual practices as designed and set up by the author for customers in Sri Lanka and for successful operation at any time. In this modern, high-tech age, perhaps some may consider these methods to be 'primitive' but they are very effective and successful because they can be used to produce good-quality foam products at highly competitive prices and can compete with large-volume producers. Especially in 'developing' countries, these practices are effectively in place.

The basics of how to plan and set up two different manufacturing plants – (a) a small-volume operation for making foam cushions, and (b) a medium-volume oper-ation for producing mattresses, cushions and sheets – are presented here for the benefit of an entrepreneur.

10.1 Example A: Manual operation

The first example is to set up a small plant to manufacture flexible PUR foam cushions for the furniture industry. All required machinery is to be designed and fabricated by the entrepreneur. Some items, such as weighing machines and a mixer, are to be purchased. This production setup will also suit a furniture manufacturer as an in-house operation.

Foam cushions: Standard size 50 × 50 × 10 cm or any other size/thickness.
Other products: Pillows, small slabs, wedges.

A building with a floor area of ≈200 m² with single-phase power and standard water supply will suffice. The building should have adequate ventilation and easy access for receiving, shipping and also to main roads. The floor area can be allocated for a small office, finished products, production/fabrication, post-curing and raw-material storage. A small space for a mini-laboratory would be very useful. If a small industrial unit is taken on lease, all facilities should be available as standard:

https://doi.org/10.1515/9783110643183-010

Figure 10.1: Foam blocks after curing and ready for cutting. Courtesy of the author's factory in Sri Lanka.

a. Building: A small industrial unit with good ventilation and basic infrastructure.
b. Machinery: A single-phase handheld standard drill with adjustable speeds of 900–1,200 rpm with a forward/reverse function would suffice as a mixer. A steel rod approximately 1.25 cm in diameter, 75 cm in length, with a 10-cm diameter disc with four flanges on one side firmly fixed to one end of the rod should be attached to the drill. The disc should be fixed with the flanges facing down. This arrangement will act as a mixer.
c. Equipment: the basic requirements would be:
    – 0–2 kg and 0–10 kg electronic weighing machines
    – 4–5 large plastic buckets with handles
    – Glass beakers and a few test tubes
    – Glass rod stirrers
    – Magnifying lens and steel rulers
    – Two drum stirrers
    – Two manual or electrical pumps
    – Two horizontal drum stands (optional)
    – One drum trolley
    – 1–2 platform trolleys
    – Miscellaneous tools

10.1.1 Raw materials

The basic components required are:
    – Polyether polyol (in drums)

- Toluene diisocyantes (TDI) (in drums)
- Methylene chloride (in small drums)
- Water
- Desmorapid OS20 or equivalent
- Desmorapid PS 207 or equivalent
- Desmorapid SO or equivalent

Only limited space is available in the factory, so it is advisable for the manufacturer to purchase the polyol and TDI in small quantities (e.g. 1–2 metric tons of each at a time), whereas the other components, being needed only in small quantities, can be purchased in the suppliers' standard eco-packs. These will last a foam manufacturer a very long time within their shelf-life periods. A sound rotating purchase system for the polyol and TDI must be ensured to prevent production downtime because of lack of these materials. If these two items are purchased by the container load, a significant price advantage can be had but, for a small producer with limited resources (and especially floor space), it is not practical.

Appropriate storage at correct temperatures is important. The polyol and the TDI must be mixed well before use. The large drums can be stacked two or four high on steel drum racks. A pump can be used for easy removal of the material by fixing a drum faucet/tap or in an upright position, if desired. The other chemicals in small packs will not pose problems because the contents can be poured out readily. Only very small quantities are required of these, so a producer may opt to transfer/fill smaller containers and use them directly for easy handling and also to minimise waste.

10.1.2 Moulds

Moulds can be made easily from wood or similar cheap materials. One can have one master, size-adjustable mould or different-sized moulds for different sizes of cushions. The latter may be the better option because different sizes of foam blocks can be made during different pouring cycles. Also, in single-sized foam block mouldings, it is advantageous to have more than one mould of the same size for increasing production volumes. Foam cushions come in 'pairs' or 'sets', so to speak, with thickness 10–15 cm for the seat cushion and 6.25–10 cm for the backrest. These are the general standards with a higher density (harder) for the seat and lower density (softer) for the backrest. Sizes may vary with different manufacturers.

Now, consider the manufacture of foam cushions $50 \times 50 \times 10$ cm (density = 29 kg/m^3) and $50 \times 50 \times 6.25$ cm (density = 21 kg/m^3). One mould can produce both sizes but different formulae have to be used to achieve the required densities. Construction of required moulds can be done by the entrepreneur.

Requirement: To produce 8 cushions from each block of foam.

To construct a simple mould, use ≈4-cm thick wood with detachable sides and bottom. The bottom plate would have a 4 (wide) × 2 cm (deep) groove in a square pattern to accommodate the vertical sides of the mould, which should fit together and be held firmly by a clamping device. The four sides sitting in the groove of the bottom plate will prevent leaking of the liquid mix. When assembled, the inner dimensions should be 51 × 51 cm and the height of the mould with allowances for the bottom (2 cm) and top (4 cm) should be ≈88 cm. Construct a simple wooden 'floating lid' of very light wood (e.g. plywood having dimensions of ≈50 × 50 cm) with a small handle on the top centre.

10.1.3 Cutting and fabrication

Conventional foam trimming/cutting is done by manual, semi-automatic or automatic band-saw cutting systems. These are very efficient computerised and accurate systems but, in some cases, a small cutting allowance must be made for the thickness of the blade. Although hot-wire cutting gives a smoother cut with less waste, some countries do not allow this practice because of the possible fire hazard. For this operation, because the foam blocks are small and to save initial capital outlay, an entrepreneur may opt for a simple hot-wire cutting system, provided the local or national laws permit the use of such a device. A big advantage is that such a device can be made easily in the factory, as described below.

10.1.4 Hot-wire cutting machine

The basic materials required to fabricate this device are:
- Single-phase 15 amp or 30 amp. Power controller 10–100 V
- Nickel/chrome: 14-G 16-G wire
- A set of electrical clips (single wire) with spring attachment
- Two lengths of ≈150 cm/5 × 1 cm aluminium channel
- A 240 × 120 × 2.5 cm warp-free board
- A steel or metal frame with legs
- A slow-speed motorised device with forward/reverse motion
- A foam shredder for foam waste
- Miscellaneous tools

Construction method: Mount the board on the steel frame base, horizontally or at 45° angle for gravity feed (if motor is not used). Cut suitable grooves in the aluminium channel to accommodate the up-and-down movement of the cutting wire attached to the clip/spring arrangement. Fix the aluminium lengths on either side of the board. Attach the two power lines coming from the power controller to the electrical clips/spring attachments and attach each wire to the two vertically fixed aluminium arms. String a suitable length of the nickel/chrome wire across the board and fix it securely and tightly.

Operation: Test the forward/reverse motion of the table. Check for motion at different slow speeds. Ensure the motion is slow and smooth, without 'jerking' motions. If satisfied, plug the power controller to a power outlet. Now, test the cutting wire by electrical heating to a 'red glow' and switch off. Repeat it once more to strengthen the wire: the device is now ready to go. Place a foam block on the board and adjust the cutting wire to the desired height and move the board forward slowly. The foam block must sit firmly on the board. By adjusting the height of the cutting wire, different thicknesses of foam cushions can be cut. To speed-up production, a second cutting machine should be available: one to cut the skins and the other to cut the final-sized cushions. By angling the cutting wire, different shapes such as wedges (angled medical slabs) and some contoured shapes can also be cut. If a smaller version of this cutting system is made with a vertical wire through a small hole in the horizontal board, contoured shapes can be cut, especially sponges. If >3 wires are used, it is best to have a 60-amp power controller.

10.1.5 Production method

a. To make a foam block of density ≈ 29 kg/m^3 to cut 8 cushions of dimension $50 \times 50 \times 10$ cm. Assuming the required moulded block size to be approximately $51 \times 51 \times 83$ cm and using the formulation shown in Table 10.1, the breakdown for the raw material components would be:

Applying, $M = V \times D$ where,

V = 51 cm $\times$ 51 cm $\times$ 83 cm
  = 215,883 cm^3
  = 0.216 cm^3
M= 0.216 $\times$ 29
  = 6.26 kg

Therefore, each batch pour = 6.26 kg + 1% waste/residue = 6.32 kg.
Note: Waste factor/gas allowance can vary from 1% to 5%.

Foaming Method: Set up the mould and line the sides with thin polyethylene film, use wax, or spray a release agent. Ensure that the liquid mix when poured will not leak from the bottom or sides. Weigh all the components accurately in different containers. Add a small quantity of colour as desired (yellow) to the polyol and mix for 60 s. Mix the water, PS 207 and the OS20 together and add to the polyol and mix for 15 s.

Table 10.1: Foam batch calculation.

Component	Formulation (kg)	Calculation	Batch quantity (kg)
Polyol	30.000	30/56.1 × 6.32	3.38
TDI	21.350	21.35/56.1 × 6.32	2.41
Water	1.825	1.825/56.1 × 6.32	0.20
Methylene chloride	2.21	2.21/56.1 × 6.32	0.25
OS20	0.560	0.560/56.1 × 6.32	0.06
PS 207	0.059	0.059/56.1 × 6.32	0.01
SO	0.061	0.061/56.1 × 6.32	0.01
Total	56.065 kg	Total	6.32

Add methylene chloride and SO to the polyol mix, and mix it for a further 15 s. Add the TDI and mix for 3–4 s and quickly pour into the mould in the liquid state before creaming starts.

If the mixing cylinder is placed inside the bottom of the mould and all the components are mixed, then it is easy to lift the cylinder and allow the liquid mixture to spread slowly and rise. The reaction starts and the foam begins to rise slowly. When it has risen to about one-third of the mould height, put the floating lid on top of the rising foam gently. This will prevent a meniscus (rounded top) being formed. Allow the foamed block to stand in the mould for about 8–10 min and, when the top is tack-free, the block can be de-moulded. In some cases, the tack-free time could be much less. Allowing the block to stand for a few more minutes after de-moulding before transporting to the final cure area enhances better foam stability. After 24 h, the foam blocks are ready for cutting and fabrication.

b. To produce a foam block of density ≈21 kg/m³ to cut 8 cushions of size 50 × 50 × 6.25 cm. On the assumption that the required block size would be 51 × 51 × 51 cm and using the following formulation, work out the component quantities as in the previous exercise.

Applying, $M = V \times D$

$$M \text{ (kg)} = \{(0.51 \times 0.51 \times 0.51) \text{ m}^3 \times 21 \text{ kg/m}^3\} = 2.79 \text{ kg}$$

Therefore, the approximate batch requirement = 2.79 + 1% waste/residue = 2.82 kg
Formulation:
- Polyol = 20.000 kg
- TDI = 16.800 kg
- Water = 1.425 kg
- MC = 3.975 kg
- OS20 = 0.400 kg

 - PS 207 = 0.048 kg
 - SO = 0.079 kg

Note: Addition of a small percentage may be considered for weight loss because of gas evaporation.

Production method: Follow the same sequence as in the previous exercise.

10.2 Example B: Making large foam blocks

This exercise will present how an entrepreneur or a medium-volume foam producer can make large flexible foam blocks using the discontinuous method in an economical manner. The production of large foam blocks necessitates more sophisticated machinery than for making just cushions, and many choices are available for a foam manufacturer. Some of the most commonly used systems are manual, semi-automatic or automatic foaming operations. Each foam block can produce almost any type of consumer foam product, such as mattresses, cushions, sheets, slabs and contoured shapes.

10.2.1 Foaming systems

PUR foaming for the producers of medium-sized volumes generally consists of a manual operation, semi-automatic or fully automatic system. Each has its own advantages and disadvantages, and the foam producer will decide what system is best for him/her.

(a) *Manual operation* involves weighing of each chemical component separately and mixing in a large cylinder that is outside or placed inside the bottom of a mould of predetermined size. In the former case, the mixture must be poured manually into the mould; in the latter case, the cylinder can be lifted by a lifting device. The volume of the mixture is large, so a powerful mixing device working at high speed should be available. The other alternative is to use a manually operated batch foaming machine (which is a compact machine). This machine has timers that can be pre-set for mixing and lifting of the mixing cylinder. The mould is brought into place under the mixing cylinder of the machine and all the components are weighed separately and poured manually into the mixing cylinder. After the completion of mixing, the bottom lid of the mixing cylinder opens automatically and the mixture pours out, spreading across the floor of the mould, and the foam start to rise. If a flat top is desired, place a floating lid on top of the foam when it has risen to about one-third of the mould height.

(b) *Semi-automatic operation*: A more sophisticated batch foaming machine is used. Different machinery manufacturers may have different features, but the general type of machines will have automatic control with timers for mixing, mixing speed, opening of the bottom lid and metering of polyol and isocyanate. This system will have two large 'day tanks' for the polyol and isocyanate that will be metered automatically to the mixing cylinder, whereas all other components are weighed separately and poured manually into the main mixing cylinder. From here the operation is as described previously. These machines will give fast moulding cycles and can produce 30–40 blocks per 8 h.

10.3 Description of a typical semi-automatic batch foaming plant

10.3.1 Foaming machine

The foaming system is based on the discontinuous method of making foam blocks one by one. The machine is a self-contained system with manual weighing and placing of all components into the mixing cylinder in a predetermined sequence. The machine is equipped with automatic mixing, opening of the bottom lid of mixing cylinder and automatic dispensing of the mixture onto the mould floor. The machine is stationary and the mould is on wheels after the foam is pulled out/removed and the cycle repeated after the 'washing mode' of the mixing cylinder. An example of basic specifications would be:
- Block-by-block production
- Foaming density 16–32 kg/m^3
- Foam block size: 2 × 1 × 1 or 2 × 2 × 1 m
- Output capacity: Up to 40 blocks per 8 h
- Electronic scales provided: 0–2 kg and 6–300 kg
- Machine space: 4 × 8 m
- Air compressor provided: 2 HP
- Catalyst weighing system: rotary stand and 10-l tanks.
- Plastic containers: one set
- Spare moulds provided: one
- Power requirements: standard three phase

Note: The mould size can vary according to the final foam block size desired but will be limited only by the shot capacity/foam volume relationship.

10.3.2 Foam cutting machines

10.3.2.1 One circular (Carousel) cutting machine
Large foam volumes have to be cut so, instead of a standard horizontal cutting machine, a more sophisticated and fast horizontal round-table rotary movement

cutting machine can be considered to be the most suitable. However, a few horizontal cutting machines are also an option.

Machine specifications/equipment:
-   Rotary slitting table with friction surface
-   Powerful vacuum system under the table to hold blocks
-   Infinitely variable table speed control (0–5 rpm)
-   Motorised slitting section
-   Tensioned Teflon-coated blade guide
-   Electric blade drive motor
-   Electric grinding attachment
-   Electric adjustment of cutting angle
-   Mode for automatic/electric push-button operation
-   Blade breakage detector with auto shutoff
-   Control panel with programmable logic control (PLC)
-   Blade 'on' warning light
-   Grinding dust exhaust extraction blower
-   One set of spares with additional blade

10.3.2.2 Vertical cutting machine

A vertical cutting machine mainly for trimming the sides of the foam blocks and can also cut sheets of thickness >1 inch.

Machine specifications:
-   Aluminium table top for free movement of foam blocks
-   Rest plate with smooth sideways movement
-   Strong steel table frame on bearing slides for easy operation
-   Slitting section with balanced blade wheels
-   Upper and lower band knife guides
-   Maximum cutting height – 120 mm
-   Maximum cutting length – 2,200 mm
-   Space required – 4.5 × 6 m
-   One set of spares

10.3.3 Foam-shredding machine

In all foam factories, the foam waste (other than what can be sold for packing) must be shredded to small pieces before it can be processed further. Unlike other plastics waste, flexible PUR foam waste is not easy to shred because of its inherent softness and flexibility. A good shredding machine (or two) employing 10-HP high speed motors with special blades are needed. The general minimal final particle size would be ≈10 mm before it can be re-bonded.

10.4 Fully automatic operation

A fully automatic operation is a very sophisticated and self-contained system. The batch-foaming machine is more or less a standard one, but has a separate control panel with touch-screen operation. Special features include storage of many formulations, pre-set automatic mixing times, mixing motor speeds, auto-metering of components and temperature controls. The system will have large chemical component tanks for the polyol and TDI with smaller tanks for all components, all connected directly to the machine for automatic dispensing into the mixing cylinder. The process is the same as in the semi-automatic case except that all operations are in sequence but fully automatic. These systems are very efficient and fast but need good maintenance to ensure the metering pumps are in good working order. The production output of these machines is much faster than other systems and would work well with a set of multiple moulds, and different densities are possible with the automatic-fed stored formulae.

10.4.1 Production (Manual or semi-automatic)

Example: To make large flexible PUR foam blocks to cut 12 twin-size mattresses (72 × 48 × 4 inches thick) of density $\approx$29 kg/m^3.

Required foam block size: 73 × 49 × 50 inches (height)

Note: Make an allowance of 0.5 inch for side skins and 1.0 inch for top and bottom. Calculating the raw material requirement of each component:

Applying formula $M = V \times D$
$V = (183 \text{ cm} \times 123 \text{ cm} \times 125 \text{ cm}) = 2.8136 \text{ m}^3$

Then, $M = (2.8136 \times 29) = 81.60$ kg

Therefore, raw material batch for pouring = 81.60 kg + 1% waste/residue = 82.41 kg
Now, calculating for each component, using Table 10.2:

Table 10.2: Foam batch calculation.

Component	Quantity (kg)	Calculation	Amount (kg)
Polyol	30.000	30/56.1 × 82.41	44.10
TDI	21.350	21.35/56.1 × 82.41	31.36
Water	1.825	1.825.56.1 × 82.41	2.70
Methylene chloride	2.210	2.21/56.1 × 82.41	3.25
OS 20	0.560	0.560/56.1 × 82.41	0.82
PS 207	0.059	0.059/56.1 × 82.41	0.09
SO	0.061	0.061/56.1 × 82.41	0.09
Total	56.065	Total	82.41

Foaming method: Prepare the mould and set it up under the mixing cylinder of the batch foamer. Meter the polyol into the mixing cylinder and add any desired colour and mix for 60 s. Add the weighed catalysts and water manually to the polyol and mix for 15 s. Add the methylene chloride and SO to the mix and mix for a further 15 s. Add TDI and mix for 4 s. All mixing times are pre-set, so the bottom lid of the mixing cylinder will open automatically after the last mix (TDI, 4 s) and the mixture (still in the liquid state) will flow across the floor of the mould and begin to rise slowly. If a flat top is desired, place a very light floating lid on top of the rising foam, when it is about one-third of the mould height.

The mould should be on wheels, and some machinery manufacturers even provide steel tracks for easy movement of the mould. Move the mould containing the foam slowly away from the batch foamer, which will now be in the cylinder 'washing mode'. The mould should be allowed to stand for ≈10 min and, when the foam is tack-free, can be de-moulded. After some time it can be transported to the final curing area. If multiple moulds are used, the production cycle can start again as soon as the washing cycle of the mixing cylinder is complete and one need not wait for de-moulding of the foamed block. Unlike in small-volume productions, the waste from the skins of these production systems form a large volume and will have to be recycled, first by particle reduction using shredders and then making large blocks using a re-bonding machine. A foam producer will have two choices – (1) steam/compression or (2) adhesive/compression. The latter is the less costly and easier method. These blocks will be ideal for use as carpet underlay, mattress bases and height enhancers for mattresses.

11 Manufacturing plants for large-volume producers

The recommended process for this manufacture is the continuous foaming method, in which large volumes of foam (>2,000 metric tons per year) can be produced. This process is very fast, so a single 8-h production shift per day is sufficient to achieve mostly large volumes. This process can make any type of flexible foam with a particular density and property on a single run or make two or three densities and properties on a single run with minimal waste at block changeovers. Unlike in the previous cases presented in this book, this manufacturing plant will need careful planning to set up a successful production unit. The capital outlay is also large, so it will be prudent to carry out an intensive project study with diligent market research and product selection, product mix and market share, as well as carrying out a feasibility analysis incorporating all these data. This study will show the best way forward, especially with the correct purchase combinations of plant, machinery, equipment and raw materials required and for effective disbursement of funds. A foam producer will initially recruit personnel with some industrial and polyurethane (PUR) foam experience and, if the project is launched on a turn-key basis, the machinery suppliers will undertake complete installation, training of personnel and initial production. The information given below is the basic guidelines for setting up such an operation.

11.1 Planned production

One could say that the entire project is dependent upon planned production. The foam producer must work out (in detail) the proposed annual volume, densities, properties and types of products. In general, in this manufacturing process, the main products will be mattresses, cushions, sheets and slabs, in which case standard machinery would suffice. However, for example, if a producer plans to market finished mattresses or contoured products such as pillows, quilting for mattresses, and rebounded blocks from waste, additional machinery and equipment will be needed.

11.2 Location and factory buildings

Once the planned foam volume has been worked out, the next item for consideration is the location of the factory. Some important factors to consider are area, local governing laws, building approvals, availability of utilities, easy access to main roads, security concerns, travelling convenience for personnel, waste disposal and access to emergency services. It may not be possible to have all factors on a 100% satisfactory basis but the important basics must be available. It is advisable to plan for possible expansion later on, so there must be extra space. It may be better to plan

https://doi.org/10.1515/9783110643183-011

for at ≥10–15 years ahead and start with a basic manufacturing operation and then gradually expand.

It is best to have separate buildings or areas for raw materials, production and final curing area, whereas one area/building can house offices, finished products/ shipping, cutting and fabrication, laboratory/quality control, small workshop/spares and recreation rooms. Although these areas are separate, they must be interconnected with easy access to each other. For safety reasons, the distances between these areas should be strictly in accordance with accepted safety standards. The floors should be of concrete ($\approx$500 kg/m^2) for industrial loadbearing, with sufficient emergency exists for evacuation of personnel and material in case of danger. Suitable fire extinguishers must be placed strategically and at least some of the key personnel must be trained in using them. Tradition and good practice is to have buildings with only ground floors (except maybe the offices).

11.3 Storage of raw materials

As in any manufacturing operation, raw material is the key factor. Special care is needed in planning storage, especially because the polyol and isocyanate present fire hazards. The floor area required will depend on the planned production volumes and production cycles. Many components are involved and even the non-availability of a single component can halt production, so the supply cycles should be worked out carefully. At this level of operation, most probably a producer will opt for bulk purchases for economic advantage and some may even prefer a just-in-time system of supplies. Although this system will save money for the producer by not holding excessive stocks, it could also create production stoppages because of delays in deliveries. Experience has shown that foam producers sometime find it difficult to secure supplies at short notice.

Some items such as tin/amine catalysts and silicone surfactants may pose a problem if excessive amounts are carried in stock because they have limited storage lives and once opened they can quickly hydrolyse and lose some of their chemical strengths. One way to circumvent this problem is to contract reliable suppliers for long periods but with regulated partial supplies in keeping with the production consumption cycles. This would allow a producer the advantage of lower costs, improved cash flow, short storage life for chemicals and the minimum requirement of storage facilities.

11.3.1 Storage conditions

If large quantities of chemicals are ordered in drums, a foam producer may have to (even temporarily) use outdoor storage. Prolonged outdoor storage should be avoided because of undesirable weather patterns with temperature variations causing condensation and possible drum corrosion. As a minimum precaution, the drums should be stored under a light roof-structure and covered. The recommended indoor storage temperature is 20–28 °C. The principle of first-in and first-out (FIFO) drums must be followed at all times to prevent the aging of chemicals.

11.3.2 Tank room

The ideal temperature of the tank room should be ≈20–25 °C and, because this is also approximately the production temperature, installation of temperature-controlling devices will be required in most cases. A suitable number of drums for several days of production should be made available in this tank room at the required temperature, ready for filling into the tanks. Whenever drums are emptied, they should be put aside (possibly in a disposal area) and replaced immediately by new full drums from the general bulk store for these drums to reach the correct temperature as soon as possible. The capacity of the day tanks is decisive for the amount of foam that can be produced with one charge. For large production runs, it may be advisable to install two tanks each for the major components (polyol and isocyanate). It is best if all chemical feed tanks are located close to the control panel and the mixing head so as to shorten the feed lines and minimise possible blockages.

11.4 Trough paper

Trough paper is an important aspect of continuous foaming. The paper rolls must be handled with care during storage and transport so as not to damage them and should be stored horizontally in a clean dry area. Standard practice is to use Kraft paper and release coated or polyethylene-lined paper of density 80–100 g/m^2. This paper is used to line the conveyor (bottom) and the sides to form a trough into which the chemical mixture will be poured for foaming. The paper roll feeding stations and the take-up stations must be in smooth working order to ensure a slow and coordinated movement of the paper trough, without stalling or the paper tearing. Appropriate feed and tensioning is of paramount importance.

11.5 General machinery

A large-volume foam producer must carefully plan all the machinery and equipment required before set up. When purchasing a continuous foaming line from a manufacturer, they will ensure that it will be complete in all aspects and will take responsibility for the installation, production trials and training of the client's personnel. However, a client must ensure that all adequate auxiliary equipment is available. The following are some of the recommended items:

- Complete continuous foaming line
- Platform weighing machine to weigh foam 'buns'
- Vertical cutting machines for foam blocks
- Horizontal cutting machines for foam blocks
- Carousel circular cutting machines
- Foam-shredding machines
- Foam-peeling machines for thin continuous sheets
- Foam waste re-bonding machine
- Edge rounding machines
- Machine for convoluted cutting of sheets
- Platform and smaller weighing machines for production floor
- Bevel cutting machines
- Profile cutting machines
- Contour cutting machines
- Punching presses

The items listed above are a range of basic machinery required on a large-volume production floor, and a foam producer will have to work out which type of machinery and the number of machines required, which will depend on the production volume and the type of products to be made.

11.6 Basic equipment

For a large-volume-foam production floor, many items are needed but the following are some of the basics required for an effective production operation:

- Forklifts
- Weighing machines
- Steel platform trolleys
- Work benches
- Steel racks
- Steel cabinets
- Shelf carts
- Hand trolleys
- Pallet trucks
- Wooden or plastics pallets
- First aid kits
- Emergency showers

- Eyewash stations
- Production boards
- Fire extinguishers
- Tools
- Drum grabbers
- Drum stackers
- Office equipment
- Laboratory equipment
- Recreation room equipment
- Safety equipment

11.7 Utilities

The content below shows the basic services required to run a large plant of this nature. Some manufacturers may opt to also have backup services, especially in the case of standby generators for electrical power. It is also wise to plan for extra installed capacities to counter future expansions:

(a) Electrical power is a key factor for a plant, so careful calculations and planning are needed. Some countries use voltages of 110/120 V, whereas others use 220/230 V and, for operating machinery, three-phase power is required. The total requirement depends on the number and types of machinery, equipment, lighting, air-conditioning and other appliances used. If all motors are to start simultaneously, more power is required than the calculated one on the rating, and it is necessary to have at least an installed capacity of >40%. Also, it may be prudent to provide for possible expansion later on. If the factory is located in an area known for frequent power failures, if resources are available, it is recommended that a suitable generator is installed as a standby.

(b) Water: No special quality requirements and ordinary industrial water supply will suffice. Although the quantity of water required for production is not great, other areas such as water installations for fire safety, personnel use, emergency showers, eyewash stations and washrooms should be given consideration. The water should be reasonably pure and, if in doubt, a water analysis should be carried out to determine whether additional treatment will be necessary. Some may even opt for ground wells and overhead tank water supplies with automatic pumping systems.

(c) Compressed air: Large factories have their own compressed air rooms by installing a several suitable air compressors and channelling supply through a large accumulator tank to the various areas of production operations. Good maintenance of these machines is important for an effective and regular smooth supply, especially to the foaming line.

11.8 Plant layout

Plant layout is an important area that needs careful planning. To achieve efficient and productive foam production, the plant layout should take top priority to ensure a smooth flow line. The operation must 'flow' from receipt and storage of goods to the production floor to the cutting and fabrication, and then again to the finished products store and shipping. All these activities must be controlled by a main office and strategically placed supervising units spread out through the entire factory.

A standard large-volume flexible PUR foam factory will consist of the following main sections:
- Storage of raw materials
- Foam production
- Foam curing
- Laboratory
- Quality control
- Small workshop (optional)
- Foam cutting and fabrication
- Foam waste recycling (re-bonding)
- Product finishing
- Finished product store/shipping
- Administration offices and personnel facilities
- Personnel training room
- Customer reception
- Security offices
- Vehicle parking
- Waste disposal

The design of the factory floor plan will depend primarily on the shape of the land block, the access routes to main roads, external utilities supply stations, production volumes planned, safety regulations and possibly on the resources available. Except for office buildings, single ground floors are recommended for the remainder of the operations.

11.9 In-house laboratory

An in-house laboratory on the production floor is a great advantage. This laboratory should have facilities to carry out material tests, density checks, formulating small foam batches with box tests, checks for cell structures and analyses of faulty productions. Three of the basic specifications required for the marketing of the foam products being produced to whatever industrial standard being adhered to are density, indentation force deflection (IFD) and fire-retardant factors. Periodic tests, reports and certifications are an essential part of a large-volume foam producer. It would be an advantage for this laboratory to have a table model electromechanical measurement machine for checking the IFD value rather than outsourcing it, which will save costs for the foam manufacturer. In highly sophisticated, very-large-volume production operations, tests are carried out even on the individual chemical components on delivery to ensure they are in good condition.

11.10 Quality control

Quality control is an essential part of any manufacturing activity. It is standard practice to manufacture any type of good to internationally accepted and recognised standards. For flexible PUR foams, a suitable and common standard is American Standards for Testing and Materials, and popular quality-controlling systems are the International Organization of Standards (ISO) series. A foam manufacturer will decide which standards to follow depending on the production/marketing aspects of the products, whether they are for local or export or both, and the regional marketplace will also play a part in this decision. Products are made to more than one standard. ISO offers some quality systems, such as statistical process control and quality assurance. These systems ensure that the foam products being made conform to the selected standards and, once these systems are in place and practised, there will be constant external audits to ensure correct practice.

11.11 Safety systems

PUR foam manufacturing deals with highly corrosive, inflammable and dangerous-to-handle chemicals, so safety (especially for factory personnel) must be given due consideration. Safety systems start with water sprinklers and fire alarms installed in strategic places, fire extinguishers (and how to use them), a spill-management emergency system in place, constant access to external professional assistance, emergency showers, eyewash stations, protective wear, sufficient emergency exits, appropriate waste-disposal methods, fire drills and emergency procedures. No-smoking signs are a must, but notices and safety posters on display on the factory floor in all areas greatly help to remind all personnel of a safety-first policy. Periodic short meetings, training updates and simulated fire-drill procedures also help to maximise safety. Some of the general causes for problems are static creating sparks, chemical spills, electrical shorts, faulty chemical formulations, bad air quality because of inadequate exhaust systems, mishandling of machinery, bad maintenance, chemical leaks and personnel not wearing appropriate protective clothing (which is also a common occurrence).

11.12 Foam production

A continuous foaming line should be installed in a separate area with the platform weighing machine at the end of the foam line. The foaming area should be well ventilated and a special exhaust system should be available to extract and take away the toxic and unpleasant fumes generated during foaming. Room temperature at processing should be at a maximum of ≈20–25 °C and, in hot climates, it may be necessary to air-condition at least the starting area where the mixing of the chemicals and foaming takes place. All chemical feed lines will be connected directly to the mixing head through the computerised control panel, and all chemical metering systems must be in good working order to prevent additional foam waste. There should be a machine-to-wall walking space of ≥2 m but much more at the end of the foaming line to accommodate the cut buns before transport to the final curing area. Some foaming operations may have motorised conveyor systems transporting the cut foam buns directly to the final curing room.

The mixing/dispensing head mixes the chemicals to predetermined formulations in sequence, and the mixture is poured onto the bottom of the paper trough, which moves forward very slowly and foaming takes place. The first 1–2 m of foam and the last ending foam (called 'end-block') has irregular shapes, an uncontrollable composition of chemicals and poses a fire hazard. For safety, these blocks may be cut and kept outside or separately until fully cured. The foaming quickly reaches its full height and width (e.g. 105×210 cm) and the slowly moving paper trough conveyor takes the continuous foam forward. By the time it reaches the cut-off saw, it is tack-free and has completed its first stage of curing. The cut-off saw can be worked manually on a horizontal traverse or electronically to pre-set lengths and cut into desired lengths of buns.

Quality control may carry out inspection of these buns and weigh all buns individually or do random weighing before they are transported to the final curing area by conveyor or manually by trolleys. Here it is important to store them ≈30-cm apart without stacking one on top of another in a well-ventilated room and they should be allowed to cure for 24 h before they can be taken for cutting. These freshly made foam blocks will have exothermic (heat-giving) reactions that can reach ≤150 °C, and time will be needed to dispense the heat and cure fully. On an approximate basis, one tonne of these foam blocks will require ≈70–80 m² of floor space. After 24-h full cure, these blocks may be stacked. Interestingly, the size of the final curing area limits the volume of the foam that can be produced at a given time because when this room is full, production has to stop for ≥24 h.

It is good practice for the production personnel or quality-control personnel to tag each freshly made block with relevant data – date, density, colour and property code – before these blocks are sent for final cure. The storing of blocks should be arranged in such a way that a FIFO system is followed to ensure that the correct blocks are taken for cutting and fabrication. If continuous thin sheeting is required for quilting in the manufacture of foam mattresses, then the foam blocks have to be round, and these can be produced by a discontinuous foaming method and cut by a peeling machine. The following are the basics of an economical continuous foaming line capable of high volumes of foam production with a density range of 12–55 kg/m³ (Table 11.1).

Table 11.1: Settings for continuous foaming machine.

Description	Machine number 1	Machine number 2
Adjustable side walls	Yes	Yes
Output	150–250 kg/min	100–200 kg/min
Number of streams	11	8
Pressure	3–5 kg/cm²	3–5 kg/cm²
Foam width	1.2–2.2 m	1.2–2.0 m
Foam height	≈1.1 m	≈1.1 m
Block cutter	Traverse type	Traverse type
Density	12–50 kg/m³	12–50 kg/m³
Polyol tank	5 tons	4 tons
Toluene diisocyante tank	3 tons	2 tons
Machine size (external)	4.4 W × 36 L × 3 H – m	4.3 W × 33 L × 3 H – m

Note: There are many different types, and the features and dimensions will vary.

11.13 Foam cutting and fabrication

The fully cured foam blocks can be brought to the conversion section by conveyor, forklift truck or manually by trolleys. The first action is to trim and remove the side, bottom and top skins of each block using vertical and horizontal cutting machines. In some advanced cutting systems, it may be possible to cut all skins simultaneously in one pass. Once this is done, the foam blocks will be ready for cutting into various thicknesses, shapes and sizes. Foaming productions are fast, so the build-up in the final curing will also be fast. To avoid cutting/conversion delays, there should be sufficient machines and operators to quickly meet predetermined targets. Here, adequate space is required, especially to accommodate additional machinery such as peeling machines and laminators if market-ready products such as mattresses, pillows, slabs and sleepers are to be made. If electronically operated semi-automatic or automatic cutting machinery are used, instead of manually operated ones, large volumes of foam can be handled without delay.

11.14 Recycling of foam waste

Manufacture of PUR foams generates foam waste mainly from the skins and at the beginning and end of a continuous production run. Other possibilities are waste

Table 11.2: Re-bonding plant specifications.

Re-bonding plant	System number 1	System number 2
Shredding machines	One – 40 HP	Two – 40 HP
Pressing system	One	Two
Hydraulic power pack	5 HP	5 HP
Mixing vessel capacity	150 kg	150 kg
Weighing box	–	One
Silo	$10\,m^3$	$10\,m^3$
Steam boiler	–	One – 80 kg/h
Blower	One – 5 HP	Two – 5 HP
Operator platform	One – 5 m	One – 7 m
Power required	70 HP	130 HP
Block size	$2 \times 2 \times 1\,m$	$2 \times 2 \times 2\,m$

because of leftovers from foam blocks after cutting and conversion as well as faulty formulating and from faulty metering during a production run. The accepted industrial process waste for foam is ≈10–15% but for large-volume productions this could be much higher. The author has had experience of some large-foam manufacturers having foam waste >30%, which is much too high, and this indicates that immediate process improvement is required. True, recycling of foam waste and sales of the products generated or the direct use of rebounded foam can contribute towards profitability but there is a limit to this activity. It is always better to set up an efficient production process with minimal waste.

Re-bonding of foam waste is carried out by first shredding the foam waste to desired small particle sizes using a powerful shredder. This material is then fed into a large silo and finally placed into a powerful pressing system (usually hydraulic) and bonded into large blocks. Re-bonded foam is suitable for mattress bases, sandwich mattresses, high loadbearing applications and carpet underlay. There are two systems that use adhesives or steam for the bonding under pressure (see Table 11.2 for examples of the two systems).

12 Recommendations for process efficiency

In manufacturing industries, it is important to have an in-house laboratory, especially for flexible polyurethane (PUR) foams because it involves chemical operations. Whether the foam producer is an entrepreneur, a medium-volume producer or a very-large-volume producer, it is essential to have a laboratory. It is not necessary to have a very sophisticated laboratory but one that can at least carry out basic tests such as density, colour, formulating, box tests, cup tests and support factor (indentation force deflection, IFD) for the larger producers. The smaller producers have the option of outsourcing any tests desired or the basics such as IFD and fire-retardant factors that customers may request as routine. Some may opt for a laboratory setup to include quality-control checks.

In planning a small laboratory, the producer may opt for using the already available expertise of one of their technical staff members. Basic important areas for consideration are:

- Location: well away from the production area
- Space: for starters, a room approximately 10×5 m with two exits
- Ventilation: good ventilation with exhaust fans is required
- Temperature control: adjustable heat control/air conditioning
- Safety equipment: fire extinguishers, spill management, protective wear, alarms

12.1 Basic laboratory equipment

Although the items mentioned earlier are simple equipment, a laboratory must have a competent qualified person to handle the tests and formulating. A chemist would be ideal but some producers may opt for one or two persons well-versed in industrial chemistry and quality control to run the laboratory. If tests such as IFD, tensile strength, tear strength, compression tests, fatigue tests and elongation are to be done in this laboratory, additional testing equipment that can be easily obtained from the many suppliers available will need to be done. However, the laboratory personnel must have a thorough knowledge and practice of using these machines properly (see Table 12.1):

For a small- or medium-volume foam producer, it may be better to outsource whatever tests are required and use these certifications as marketing tools. For the large-volume producer, a good option would be to install at least a table model electromechanical testing machine for IFD (support factor) tests as this is essential for marketing. With some modern testing machines, it is possible to carry out more than one test on the same machine. A producer must use some internationally recognised standards for successful marketing of foam products. In the western world probably the most popular one is the American Society for Testing and

https://doi.org/10.1515/9783110643183-012

Table 12.1: Laboratory equipment.

Description	Quantity
0–5 kg electronic weighing machine	1
0–10 kg electronic weighing machine	1
Optional items	Chemical balance, Bunsen burner
Industrial thermometers	3
Burettes	2
Pipettes	2
Small beakers	5
Medium-sized beakers	5
Glass stirrer rods	4
Electronic clocks with second hands	2
Stopwatches	2
Small mixers 900–1,200 rpm	2 (single-phase/handheld)
Small high-speed mixer 3,000 rpm	1
Small wooden boxes	4 (detachable sides: 30 × 15 × 15 cm)
Small hot-wire cutting system	1 (single phase/height adjustable)
Small chemical fire extinguishers	2
Sharp slicing knives	4
Plastic buckets – different sizes	10
Safety equipment safety wear	Glasses, boots, masks, absorbents
Miscellaneous items	Gloves, lab tools, scissors, magnifier
Other items	Pens, paper, calculators, rulers
General items	Desks, chairs, laboratory stools, phones
Sink with water supply	1

Materials. Laboratory coats are an option but the provision of a gas/carbon dioxide monitor enables periodic monitoring of the atmosphere in a factory to ensure the toxic vapour levels are within the acceptable limits.

12.2 Recommendations for the efficiency of foam plants

PUR foam is a booming industry, with the demand for flexible foams increasing rapidly, with population growth, all over the world. Setting up a foam plant is a factor and the efficient running of it is another matter, necessitating close supervision and correct decision-making (as in all manufacturing operations). Plant efficiency may range from 60% to 80% with an efficient plant close to the upper limit. The following parameters discussed are useful for achieving a high level of efficiency.

12.1.1 Key factors on a production floor

Many factors influence the efficient operation on a factory floor. To achieve this, the following are recommended:
- Mission statement: company objectives, company vision, customer policy
- Posters: motivation, safety, warnings, hazards, accidents
- Machinery: proper installation, operator training, maintenance, start-up and shut-down procedures
- Maintenance: routine and preventive maintenance, spares
- Quality assurance: quality control, documentation, analysis, reporting, action
- Process engineering: time and motion study, constant process improvement
- Product development: quality improvement, new products, cost-cutting
- Production methods: operating systems, procedure, correct tools
- Workforce: good morale, technical ability, good attendance, teamwork
- Productivity: achieving targets, input/output ratio
- Raw materials: material quality, storing, contamination, correct formulating
- Ventilation: adequate ventilation, efficient exhaust systems, air quality
- Good management: communication, motivation, evaluation, incentives
- Documentation: recording systems, reporting, feedback, action
- Warehouse: efficient operation, storing, shipping
- Troubleshooting: solving production problems, spill management, returns
- Fire precautions: fire extinguishers, training, emergency exists, fire drills
- Employee facilities: lunch room, washrooms, lockers, uniforms, parking
- Tools and equipment: proper training, inventory and issue systems, safety issues
- Production floor committee: manager, supervisors, lead-hands, periodic meetings
- Downtime: minimise downtime, a standby generator where practical
- Constant checking: plant manager/supervisors, lead-hands
- Operating difficulties: supervisors, efficient reporting, finding solutions
- Logistics: proper flow in all areas
- Job rotation: operator rotation between hard/easy jobs
- Absenteeism: correct selection of personnel, attendance bonus
- Rivalries: avoid rivalries between operators/supervisors

12.2.2 Preventive maintenance in foam plants

The efficiency of a foam plant will depend greatly on a planned preventive maintenance system and how well it is implemented and carried out. A small in-house workshop and a good inventory of spares (especially of parts hard to purchase) and well-trained personnel will ensure smooth production operations with minimal downtime.

Some basic recommendations are detailed as follows:

- Make sure that the chemicals in all the drums or storage tanks are stirred at least once every other day for 10 min. This will ensure that the chemicals do not settle and collect moisture, and circulation will ensure that all valves and pumps do not seize up and lead to system failure. If fillers are incorporated in blended polyols, this becomes critical because the filler powder will settle on the pump shafts and also make the valves stick. Pumps need to be re-calibrated every week or so to ensure that pump output is consistent.
- A daily check of the foam block cutter operation is desired. This should include:
- A check of the cutting blade for wear. It should be free of any accumulated foam build-up. The guide rollers should be clean and move freely.
- A check of the spider coupling for the conveyor drive motor should be done for wear and damage.
- An operational check in the full automatic mode to ensure cutter runs operates as it should.
- A daily check of the slate conveyor for damaged slates, which may cause the conveyor to stick during operation. The guide rollers and connecting links should also be checked for proper positioning because this may also lead to sticking during operations. The pipe conveyor should be checked for security and operation of the slip-clutch, as well as all drive chains for wear, damage and proper installation. Check whether all safety guards are in place.
- All fluid lines should be changed once annually. Even though lines may appear to be in good condition on the outside of the line, internally they may be clogged, restricting the flow of chemicals or air, and may lead to premature failure and poor foam quality.
- An oil-level check in all motor gear boxes to be done at least once every 6 months to avoid premature failure of pumps, gearboxes and motors.
- The mixer head assembly should be overhauled every 2,000 blocks or once every 2 months. This is to avoid seal failure leading to foam leakages during operations.
- The isocyanate valve should be overhauled every 800 blocks or once a month. The internal components of their valves may become partially blocked and restrict the chemical flow or hinder full valve operation. This may not be noticeable and can lead to poor foam quality.
- At the completion of each pour, run the mixing head at approximately 150–200 rpm and allow polyol to flow through the mixing head until it runs freely and cleanly. This will prevent clogging of the mixing head.
- At least once a week, the main shaft of the mixer head should be lubricated. Grease should be pumped into shaft bearings housing until clean grease is seen coming out of either end of the housing.

In various manufacturing industries, there may be different standards for efficiency but, in general, an efficiency factor of ≥80% can be considered good.

Foam manufacturing operations are not easy, especially if large volumes are involved and a large-volume foam producer may feel even a lesser efficiency factor is reasonable.

12.3 Guidelines for a quality control system

An inherent problem with PUR foam production is the generation of waste. This, if not controlled effectively, can lead to excessive waste as experienced by the author in some large-volume foam producers, where the foam wastes were 30–35%, which is excessive and unacceptable. A reasonable waste level may be considered to be 15–20%, with the lower level being preferred. Then, there is the most important factor of foam quality to be considered. From a marketing angle, the two most important factors are density and the IFD, which the average customer will look for. To achieve these required marketing standards, many basic factors have to be controlled: raw materials, foam block weights, density uniformity, dimensions of blocks and products, and aesthetic aspects. Every production floor needs an effective quality control system and this chapter presents the basic guidelines and explanation of a suitable system: Statistical Process Control (SPC) – International Organization of Standards (ISO) 9000 series.
SPC is a specific tool used in quality control to:
– Achieve a quality product/service
– Maintain the quality of a product/service
– Improve the quality of a product/service

SPC basics are as follows:
– Mission statement: company policy, customer policy, quality assurance
– Organization chart: company structure, lines of authority, reporting systems
– Quality system implementation: in-house or external professional expertise
– Implementation methodology: procedure
    – Training of personnel
    – Monitoring and recording
    – Evaluation
    – Corrective action (if necessary)
    – Company audits for compliance
    – Customer feedback
    – ISO audits/continuing certification

Once the SPC quality system is running, constant evaluation and action is needed on a production floor at all levels to ensure that procedures laid down are being followed. Maintaining quality is an ongoing operation and needs the cooperation of all.

SPC is only a tool for quality maintenance and constant improvements starting from raw materials to the final finished products. Quality is not the responsibility of a person or a department, it is everyone's job; in other words, 'teamwork'.

Quality control + quality management = quality assurance

SPC uses two basic control charts called the X-R chart and the P chart, as discussed later.

12.3.1 What is a control chart?

A control chart is a graphical representation of the quality of a particular feature of a part or product (weight, density, dimensions or other) generated by a process.

Why use a control chart? Once documented and analysed or while a process is in progress, it tells 'how well' the process is performing to pre-set tolerances and what action is needed to keep it within the limits.

The X-R control chart has four basic sections:
- Section one: the title box tells us
  - The name of customer
  - What is the product
  - Which feature is being measured
  - How often it is done
- Section two is the graph (or grid) for charting the averages.
- Section three shows the areas where operators fill in the following data:
  - Date
  - Time
  - Measured values
  - Sum total
  - Averages
  - Range
  - Initials
- Section four is the graph for plotting (charting) the range.

12.3.2 What is an X-R chart?

An X-R chart is a chart used for recording variable data.
  For example:
- Measuring a feature of a product
- Quantitative (length, density, size or other)

12.3.3 What is a P chart?

A 'P Chart' is used for attribute data. This exercise is to determine the good or bad, accept or reject concept of a product/products. For example, in foam, attribute data are quality, density, colour, dimensions or others. While an X-R chart plots the course of variations in a process in progress, a P chart charts/records the concept of accept/ reject of foam products in the final stage.

Every process has some kind of variation and it may be true to say that no two products may be exactly the same but can be accepted if they are within the predetermined tolerances. The variations between products may be very small but sensitive and accurate instruments will be able to detect these variations. The variations detected are what are recorded in the control charts for necessary action. In a process, there are two main types of variations. First, they are random or common. These variations are due solely to chance: humidity, temperature or voltage fluctuations, material qualities and equipment performance. These variations are more or less uncontrollable and difficult to eliminate. Second, there are assignable variations that are due to 'specific findable causes' that can be rectified, such as faulty formulating, incorrect metering settings, wrong chemical weighing and operator errors. These are the main reasons for making a process go 'out of control'. If a plotted or charted point is out of the upper control limit (UCL) or the lower control limit the process is said to be out of control. If all the plotted or charted points are within the control limit lines, then a process is said to be 'under control'. If the plotted or charted points are very close to the actual desired standard (central horizontal line), the process is in excellent shape and can be deemed to be working at the best level.

After an ISO system (9001, 9002, etc.) has been selected, a business must decide on the correct procedure to implement it. Management with the quality control personnel will have to work out the areas of operations where this system should be put into practice to obtain effective results. Tolerances may depend on customer's specifications or what foam standards are followed, or it may well be that a foam manufacturer will have his/her own standards.

Effective initial training is a must for all personnel involved to clearly understand what is expected. In practice, there are bound to be various problems such as operator inadequacy, lack of understanding, faulty recording and reporting. To counter this, periodic updating training sessions and extra checking of the system activity in all areas by supervisors will help greatly. Videos, handouts and discussions during continued training will be very effective.

This is initially done by operators on a job. Supervisors and lead-hands should make sure that the quality control system documents are available at each station. A good rapport between the operators and the floor supervisors will ensure efficient and accurate work.

In a large-volume foam manufacturing operation, there would be a quality controller with a team of quality inspectors working the production floor. They will be constantly checking the ongoing activity and ensure smooth and correct reporting procedures. It is the responsibility of the floor supervisors to take corrective action if problems are reported by operators or the inspectors to keep any process in control. All system documents should flow back to the quality department for evaluation and action under the supervision of the controller.

Corrective action is very important and cannot be done haphazardly or postponed. There are bound to be problems on the production floor, so the supervisors must be alert at all times to take corrective action, however small the problem may be. Especially in foam cutting, fabrication and assembly, if faults are not detected instantly, several days could elapse before a quality problem is detected, resulting in appreciable loss for a business.

Periodic company audits help greatly for maintaining quality assurance in a factory. The ideal would be for an independent team (staff other than quality control personnel) or an external management team to carry out these audits for conformity to predetermined standards. Some factories may even have a small team led by a lead-hand of one department carrying out the audit of another. It is an accepted concept that an independent body would 'see' problems and inaccuracies better than the same people who deal with them on a regular basis.

An effective customer feedback system is an essential part of any quality programme. The four basic divisions involved in the operation are production, quality control, shipping and marketing. A constant and free-flow dialogue/ reporting system between marketing and management personnel with customer feedback will help to maintain quality at all levels. In general, a marketing team would build up a good rapport with customers and most would overlook minor quality problems. However, top-level industrial customers such as automotive or medical application customers would be stricter in their quality requirements. For a foam manufacturer, some of the areas where quality problems may arise are support factor, non-uniform density, product dimensions, aesthetic values or packaging.

A business wanting ISO certification will seek the advice and guidance of the quality management systems representing ISO in that province or country. Once this is done they will study the foaming operation and work out a system suitable for that operation. They will provide the basics such as a quality system manual, forms, training and implementation. Once they are satisfied, the business will be certified to whatever standard decided on. To maintain valid certification, the ISO branch will send their officers for periodic audits to ensure efficient operations to maintain quality at all levels. An ISO certification is a big boost and, in some sectors, a must for successful marketing.

12.4 General recommendations for troubleshooting

In foam production, whatever the process, there are always product deficiencies and the following are recommendations for correcting these problems. Table 12.2 shows solutions for possible problems for the discontinuous process, whereas Table 12.3 shows solutions for moulded foams. Table 12.4 highlights possible processing faults and possible solutions for slabstock foaming using the continuous process.

Table 12.2: Recommendations for foaming troubleshooting.

Problem	Possible causes	Corrective action
MDI too thick	Low MDI temperature Temperature damage	Increase to correct level Contact supplier
MDI gelled	Extreme temperature damage	Contact supplier
Skin on MDI	Moisture contamination	Filter and transfer to new container
Solids in MDI	Frozen MDI Moisture contamination Temperature damage	Increase temperature Filer and transfer to new container Contact supplier
Polyol too thick	Low temperature	Increase temperature to correct level

MDI, methylenediphenyl diisocyanate.

Table 12.3: Cup shot test.

Problem	Possible causes	Corrective action
Reaction too fast	Material temperature high Wrong ratio Chemical problems	Adjust temperature and redo test Check and redo test Contact supplier
Reaction too slow	Material temperature low Wrong ratio Poor mixing Chemical problems	Adjust temperature and redo test Check and redo test Check stirrer/increase time Contact supplier
Free rise density high	Lack/loss of blowing agent Wrong ratio Poor mixing	Check water content Check and redo test Increase speed/time
Free rise density low	Excess water in polyol	Check/contact supplier

Table 12.4: Deficiencies in slabstock processing.

Defect	Description	Recommendations
Bottom cavitation	Bottom eaten away	Look for errors in metering Decrease tin catalyst
Dense bottom skin	Denser foam at bottom	Increase silicone level
Creeping cream line	Cream line moves back	Speed-up conveyor Increase conveyor angle Lower amine catalyst level
Smoking	Excessive isocyanate vapours/errors in metering	Reduce isocyanate level
Tacky bun surface	Foam surface sticky too long	Increase total catalyst levels Look for errors in metering
Flashing/sparklers	Effervescence on rising foam	Decrease isocyanate Decrease silicone/amine Increase tin catalyst Decrease component temperatures
Friable skin	Skin flakes off at touch	Increase/change amine catalyst Increase component temperatures
Gross splits	Vertical/horizontal splits	Increase tin catalyst Decrease amine catalyst water percentage Increase silicone
Heavy skin	Thick, high-density skin	Increase all catalysts Increase isocyanate content
Moon craters	Small pockmarks on buns	Reduce air entrapment in pour Minimise splashing on pour
Pee holes	Small spherical holes	Increase silicone Decrease mixing speed Decrease tank agitation Reduce air entrapment in pour
Relaxation	Foam rises and settles down	Check/increase tin catalyst Check/increase silicone level Reduce amine catalyst level Reduce mixing speed/nucleation
Shrinkage	Bun contracts on cooling	Decrease tin catalyst/silicone Decrease isocyanate index Increase mixing speed Check air injection

Table 12.4 (Continued)

Defect	Description	Recommendations
Stratification	Irregular density throughout	Look for metering errors
Undercutting	Reactants flow under rising foam	Speed-up conveyor Decrease conveyor angle Increase catalyst levels
Zigzag splits	Crumbly ragged splits	Increase tin catalyst
Boiling	Large bubbles appear and burst	Check tin/silicone quality Reduce amine catalyst
Crazy balls	Liquid bubbles on surface	Increase mixing speed Minimise splashing on pour
Odour	Objectionable odour of foam	Use amines with less odour Allow more time for degas

Appendix 1

Useful conversion tables

The following conversion factors dealing with the metric system of units are presented for the convenience of readers.

Base units: Length   Metre (m)
            Mass   Kilogramme (kg)
            Time   Second (s)
            Electric current   Ampere (amp)
            Thermodynamic temperature   Kelvin
            Amount of substance   Mole
            Luminous intensity   Candela

Conversion Table of Units Commonly Used in Polyurethanes

	To convert	Intro	Multiply by:
Airflow	ft^3/min	litre/s	0.4719
	ft^3/min	m^3/s	0.0004–719
Coating	g/m^2	ounces/yd^2	0.0295
	ounces/yd^2	g/m^2	33.90
Density	lb/ft^3 (pcf)	kg/m^3	16.018
	kg/m^3	pcf	0.0624
	pcf	$g/cc(g/cm^3)$	0.016
	g/cc	pcf	62.43
	g/L	pcf	0.0624
Energy	Joule	Ft-lb	0.7573
	Joule	In-lb	8.85
	ft-lb	Joule	1.355
	in-lb	Joule	0.113
	Btu	Joule	1.055×10^3
ILD or IFD(LOADBEARING)	$lb/50\ in^2$	$N/323\ cm^2$	4.448
	$N/323\ cm^2$	$lb/50\ in^2$	0.225
	$lb/50\ in^2$	$kg/323\ cm^2$	0.4536
	$kg/323\ cm^2$	$lb/50\ in^2$	2.2

Figure A1.1: Conversion table. Adapted from information given in Acknowledgements.

https://doi.org/10.1515/9783110643183-013

	To convert	**Intro**	**Multiply by:**
Length	Angstroms	Meters	1×10^{-10}
	Metres	Micro	1×10^{6}
	Micron	Angstroms	1×10^{4}
Pressure and Stress	lb/in^2 (psi)	(kPa)kN/m^2	6.895
	kN/m^2 (or kPa)	lb/in^2	0.145
	kg/cm^2	kPa	98.07
	kg/cm^2	psi	14.223
	psi	kg/cm^2	0.0703
	psi	Pa	6,895
	Pa	psi	0.000145
	g/cm^2	psi	0.01422
	psi	dyne/cm^2	68,965
	dyne/cm^2	psi	0.0000145
	bar	atm	0.987
	bar	kg/cm^2	1.02
	bar	Pa	1×10^{5}
	dyne/cm^2	atm	9.869×10^{-7}
	psi	atm	0.068
	Pa	dyne/cm^2	10
	MPa	psi	145
	atm	Pa	1.013×10^{5}
	dyne/cm^2	Pa	0.1
Tear Strength	lb/in	N/cm	1.75
	N/cm	lb/in	0.571
	N/m	lb/in	0.00571
	lb/in	N/m	175.1
	lb/in	kg/cm	0.1786
	kg/cm	lb/in	5.6
	kN/m	N/cm	10
	kg/cm	N/cm	9.798
Temperature	°C	°F	9/5 (°C) + 32
	°F	°C	(5/9)(°F −32)
	°C	°K	°C + 273.15

Figure A1.1 (continued)

	To convert	Intro	Multiply by:
Volume	Litre	in^3	61.023
	Litre	U.S. gal	0.264
	Litre	ft^3	0.0353
	U.S. fl oz	mL	29.6
	U.S. fl oz	m^3	2.957×10^{-5}
	U.S. gallons	Litres	3.79
	Litres	m^3	0.001
	m^3	Litres	1,000
	Litre	U.S. fl oz	33.819
	U.S. fl oz	British fl oz	1.0408
	Board foot	m^3	2.359×10^{-3}
Viscosity	Centipoise = centistokes × density		
	Pascal second(Pa's)	Centipoise	1,000
	Centipoise	Pa's	0.001
	mPa's	Centipoise	1
Weight	g	Metrictonnes	1×10^{-6}
	g	kg	0.001
	g	oz (avdp)	0.0352739
	g	Pounds (avdp)	0.0022026
	Ounces (avdp)	g	28.3495
	Ounces (avdp)	Metric tonne	2.83495×10^{-5}
	Metric tonne	kg	1,000
	Metric tonne	Pounds	2,240

Figure A1.1 (continued)

Appendix 2

Suppliers of raw materials, foaming and cutting machines

North American, European and Asian manufacturers and suppliers of polyurethane foam raw materials, foaming machinery, cutting and conversion equipment service the flexible polyurethane industry. Some of them are detailed below.

A2.1 Raw materials

- Dow Corporation, USA
- BASF AG, Germany
- Bayer AG, Germany
- Hunstman Corporation, USA
- Chemcontrol Limited, USA
- Issac Industries Incorporated, USA
- Era Polymers Limited, Australia
- Union Carbide (Canada) Limited, Canada
- Premilec Incorporated, Quebec/Canada
- Bio Based Technologies LLC., USA

A2.2 Foaming machinery

- Hennecke Gmbh, Germany
- Canon Viking, UK
- Beamech Limited, UK
- Laader Berg AS, Norway
- AS Enterprises, India
- Modern Enterprises, India
- Edge Sweets Company, USA
- Elitecore Machinery Manufacturers Limited, China
- Sunkist Chemical Machinery Company Limited, Taiwan
- Dongguan Hengsheng Machinery Company Limited, China

A2.3 Cutting and fabricating machinery

- Baumer AG, Germany/USA
- Edge Sweets Company, USA
- AS Enterprises, India
- Modern Enterprises, India

- Elitecore Machinery Manufacturing Limited, China
- Croma Foam Cutters Incorporated, USA
- Wintech Engineering Limited, Australia
- Demand Products Incorporated, USA
- Sunkist Chemical Machinery Limited, Taiwan

Abbreviations

AC	Alternating current
CAD	Computer aided design
CAM	Computer aided machining
CFD	Compression force deflection
DNA	Deoxyribonucleic acid
FIFO	First-in, first-out
FPF	Acronym for flexible polyurethane foam
GPM	Gross profit margin
HR	High resiliency
IFD	Indentation force deflection
ISO	International Organization of Standards
LCL	Lower control limit
MDI	diphenylmethane diisocyanante
MSDS	Materials safety data sheets
NCO	Isocyanate reactive group
NMR	Nuclear magnetic resonance
PLC	Programmable logic control
PU	Polyurethane
PUR	Polyurethane
PVC	Polyvinyl chloride
QC	Quality control
ROE	Return on equity
ROI	Return on investment
SPC	Statistical Process Control
TDI	Toluene diisocyanate
UCL	Upper control limit
UV	Ultraviolet

https://doi.org/10.1515/9783110643183-014

Glossary of terms

Auxiliary blowing agent an additive used in the production of foams which supplements the primary blowing agent (water) to make foam softer or lighter.
Accelerator chemical substance (catalyst) to increase the rate of chemical reactions.
Acid number value given to trace residues of acids in finished polyols.
Additive a substance added to a foam mixture to enhance or achieve a property.
Adjustable crossbar device in continuous foaming machines to adjust the height of the foam mix.
Adjustable speeds ability of pumps and mixers of machines to work at different speeds.
After-cure period of time after all chemical reactions have ceased.
Agitators mechanisms that mix chemical components in tanks before mixing together.
Aliphatic division of organic compounds containing carbon with open-chain structures.
Ambient normal temperature of a room or environment.
Amine catalyst compound used to accelerate the blowing reaction.
Anisotropic foam having different properties along axes in different directions.
Antioxidants materials added to foam to resist oxidation and improve properties.
Antistatic agents chemicals added to prevent accumulation of electrostatic charges.
Aromatic organic compounds having a benzene ring-type molecular structure.
Air flow a measure of the ease with which air passes through a foam sample.
Acoustic foam foam designed to dampen vibrations or sound.
Ageing accelerated tests to determine changes in foam properties over time.
Auto solvent flush system for automatically cleaning the foam mixing head.

Ball rebound a test set by the American Standards of Testing and Materials to measure the surface resilience of foam.
Board foot a unit of measurement for flexible foams – square foot by one-inch thickness.
Bonded foam shredded foam scrap bonded. Used as mattress bases and carpet underlay.
Bun a segment of foam cut off from continuously produced slab stock.
Batch mixer a machine with a dispensing head in which all chemicals are mixed and poured.
Batch mixing a process in which all measured chemicals are mixed in one container.
Batch tank large containers in which 2–3 chemicals are mixed separately.
Belt conveyor a strong metal belt lined with paper to convey the mixed foam.
Blanket heater provides heat at desired levels for tanks containing chemicals.
Block a cut-off foam block from a continuous production line or singly made block.
Blowing agent a chemical compound that provides gas for foam expansion.
Blowing reaction chemical reaction resulting in the release of carbon dioxide.
Boardiness foam that feels stiff and not soft, yet somewhat flexible.
Boiling gas generated during foaming not contained inside and which is escaping.
Bonding synonym for gluing, laminating or re-bonding.
Bones slang term for high-density streaks or flow lines on the bottom of foam blocks.
Bulking agent material added to increase the volume of a foam mixture.
Bulk storage large tanks or containers to store components in large volumes.
Burp condition when a puff of gas is released during foaming.
Block density practice of determining the density of uncut foam blocks for process analyses.

Cell open space/cavity in flexible foam surrounded by polymer membranes.
Cell opening breaking of polymer membranes to form open cells.
Closed cells unopened polymer membranes in foam structure.
Combustion modified foam flexible foam made with additives to reduce ignition.
Comfort factor degree of deflection on the foam surface to minimise pressure.

https://doi.org/10.1515/9783110643183-015

Compression loss of initial height of foam because of body weight.
Conventional flexible polyurethane foam flexible foam made from polyether types by basic methods.
Convoluted flexible foam sheets fabricated to have peaks and valleys.
Cradling ability of foam cushions to distribute body weight uniformly.
Calibration synonym for controlled metering of components into the dispensing head.
Catalysts compounds capable of increasing the speed of chemical reactions.
Catalyst balance the ratio between the amine and tin catalysts in foam formulations.
Catalyst mix a combination of catalysts to help balance blowing/gelation reactions.
Cell Structure cavities left in foam after the cell walls have polymerised completely.
Cell count the number of cells per linear inch or centimetre.
Cell membrane thin intact polymer film that forms a cell.
Cell size average diameter of the pores, often referred to as fine, medium or coarse.
Cellular plastic expanded or foamed plastic such as expanded polystyrene.
Centipoise 1/100th of a poise. Viscosity is commonly expressed in poise.
Chain extension end-to-end lengthening of main polymer molecular chains.
Channelling a small-scale or narrow undercutting of rising foam.
Charging describes the filling of machine tanks with chemicals.
Chiller method of maintaining chemical components at desired temperatures.
Chopper device used to cut-off foam blocks or to cut foam trims to small pieces.
Clickable foam foam that recovers 100% from pinching effects.
Closed cells completely closed foam cells, as in the case of semi-rigid or rigid foam.
Closed moulding process of moulding a foam object (e.g. a foam pillow).
Cold moulding process of flexible foam moulds at <85 °C.
Collapse the sudden or complete loss of foam height after the foam has risen.
Colourants dyes or pigments used to colour foam.
Complete package term commonly used for a complete continuous foaming line.
Components liquid chemicals separately metered onto a main foaming stream.
Compression load deflection resistance to compression of a sample of flexible foam.
Compressive strength generally refers to the resistance of rigid foam to compression.
Continuous mixing head mixing device such as on a continuous foaming line.
Continuous slab stock production of continuous uniform loafs of flexible urethane foam.
Conventional system method of foam production in which a mixing head delivers liquids.
Converting common term for the cutting and fabrication of foam as saleable items.
Conveyor belt moving loop of material (metal, rubber or paper) to convey foam.
Convoluted cutting splitting of a flexible foam sheet to get surfaces with peaks/valleys.
Core density density at the centre of a foam block/object.
Cracks long narrow openings or cavities because of faulty formulation or weighing.
Cratering 'moon cratering' on foam block because of the opening of gas bubbles.
Crazy balls globules rising from within the foam to the surface.
Cream time the time between pouring and the point at which the liquid mix turns creamy and starts to expand.
Cross-linking setting up of chemical links between polymer molecular chains.
Crown flexible foam mattress or cushion thicker in the middle than at any edge.
Crumbs very fine pieces of flexible foam from shredding, milling or ground scrap.
Cubic centimetre measure of volume in the metric system.
Cubic metre cubic measurement where the length, width and height are all one metre.
Cure a term which denotes the completion of a 100% chemical reaction.
Curing time the time required to achieve the desired level of curing.
Cut-off saw foam-cutting device to make vertical cuts on a continuous foaming line.

Density a measurement of mass per unit volume lbs/ft^3 or kg/m^3.
Durability denotes how well a flexible foam retains its comfort, support and shape with use.
Dynamic fatigue a durability test in the laboratory using roller-shear or pounding-type mechanisms.
Day tanks expression meaning batch or machine tanks for daily production.
Degree of polymerisation number of structural units ('mers') in the average polymer molecule.
De-moulding time time between pouring the liquid mix to the removal of the foamed article.
Die cutting a process in which shaped foam articles are cut out from a sheet of foam.
Differential thermal analysis a test to assess the behaviour of foams exposed to high temperatures.
Diisocyanate a major chemical component used in the production of flexible foams.
Dilantic term applied to a liquid that resists being moved but is quite fluidic when at rest.
Diluent a material used to extend or bulk-up another material without changing reactivity.
Dimensional stability ability of a plastic part or product to retain its original shape.
Dimer a substance comprising molecules formed from two molecules of a monomer.
Diol a polyol or resinous material having two hydroxyl groups attached to each molecule.
Discharge orifice the opening through which the chemical mix is discharged.
Discolouration the gradual yellowing of urethane foam because of exposure to ultraviolet light.
Distribution block manifold on continuous foaming used for collection of a stream of chemicals.
Dribble marks long lines of undesirable bubbles, usually just under slab stock skin.
Dry heat aging an accelerated aging test for formulation materials.
Dynamic balance term used to describe stability achieved on operating variables.

Elastic component viscosity term for resistance of some fluids to flowing under shear and
    applied force.
Elastic limit the point of deformation beyond which a material will deform permanently.
Elastic modulus the ratio between the force applied to cause deformation and the resistance to
    that force by the material.
Elastomer a rubber-like material, but not necessarily made of rubber.
Elongation the percentage of its original length to which a specially shaped sample will stretch
    before breaking.
Emulsifiers additives to a formulation that help in stabilising a mixture between the time of
    mixing and pouring.
Emulsion a suspension of fine droplets of one liquid in another.
Equivalent weight the molecular weight of a chemical divided by the number of functional groups.
Exothermic heat giving/generating chemical reaction such as the urethane foam reaction.

Flex fatigue the loss of flexibility of foam after flexing a pre-determined number of cycles.
Fast heat intense heat supplied to a mould to improve skin quality and the physical properties of
    a moulded flexible foam product.
Fast heat oven an oven capable of supplying the intense heat required for the 'fast heat' process.
Filler an inert material added to a foam formulation to reduce costs/change physical properties.
Filter a device used to remove unwanted particles from liquid chemical streams.
Fine cells a term used to describe foam cells having a cell count of >80 per linear inch.
Fixed calibration time process of metering chemical liquids by a time limit, usually in seconds.
Fixed ratio the condition of having all components with a fixed throughput without variations.
Flak a slang term used to describe many small splits scattered widely.
Flame lamination process of sticking together foam and other material by heat/melting.
Flame retardant foam material imparting a certain degree of resistance and slowing down burning.
Flammability degree of burning in various given situations.
Flaps usually semicircular internal blows or voids on the skin surface and also connected inside.

Flashing a condition in the continuous foaming process in which continual release of tiny bubbles of gas from the surface occurs.

Flex fatigue the loss of physical properties of a foam sample undergoing continuous flexing.

Floating lid a special lightweight lid used in foam production to prevent the top from curving.

Flow lines pattern of high-density streaks or ridges radiating from the bottom of a foam block.

Flowmeter device used to gauge the flow of liquids.

Flow rate refers to the quantity of chemicals delivered to the metering port.

Foam term used to describe material produced with a gas reaction with open or closed cells.

Formula chemical components for foam production.

Formulation working out correct quantities of chemicals to achieve desired results.

Free rise unrestricted rise of reacting foam.

Frothing practice of incorporating an unusually low boiling material into a foam mixture.

Functionality the number of reactive groups attached to a single molecule.

Gas loss net weight loss between chemicals used and final foam weight.

Gel strength denotes the stability of the foaming mass.

Gelation rate the speed with which the chain extensions and cross-linking reactions occur in the foaming mass.

Gelling reaction the increase in viscosity of the foaming mass caused by polymerisation of the liquid chemicals.

Glycol generic term for polyols having a functionality of 2 (also termed 'diols').

Grinders machines that produce small foam particles from foam trims or scrap. These machines are also called shredders, cutters and hammer mills.

Hand batching the practice of weighing all ingredients by hand separately.

Hard spots high-density areas formed in a moulded foam product.

Haze smoke or fumes released from the foaming mass during reactions.

Health bubbles little bubbles popping up through the top skin of a foam block, and generally accepted to be a foam of good quality.

High-resilience flexible polyurethane foam flexible foams having a high support factor and greater surface resilience than conventional foam. Also has less uniform, more random cell structures.

Hysteresis the ability of a flexible foam to maintain its original characteristics after constant load.

Heavy skin a very thick outer skin in free rise or moulded foam.

Helix mixer term describing a mixer with a helical spiral mixing technique.

High shear mixer a mixer with a blade or impeller that transfers a large amount of mechanical energy to a chemical mix in the form of heat.

Holding tanks storage tanks in which stable premixes are made and stored before use.

Horizontal cutting practice of cutting foam blocks with the blade parallel to the foam surface.

Hot wire cutting cutting foam by means of an electrically heated wire.

Humid aging accelerated aging test for foam under conditions of high humidity and temperature.

Hydraulic a system by which energy is transferred from one place to another by compression and the flow of fluids.

Hydraulic traverse a spreading device that can be used on a continuous foam production line.

Hydraulic mixing head a mixing device in which the primary mixing is the turbulence created by the flowing streams of chemical components.

Hydrolytic stability ability of a foam to withstand hydrolysis or disassociation with water under constant exposure.

Hydrophobic water repellent.

Hydrophilic affinity for water. Foams with large open cells are more absorptive sponges.

Hydroxyl group the combined oxygen and hydrogen radical that forms the reactive groups on polyols.

Indentation force deflection value indentation force deflection value formerly called rubber Manufacturer's Association value. Taken as a compression value of 25%.

Indentation force deflection index indicates support value of a foam based on compression. This value is the ratio of 65% compression value divided by 25% compression value. Index values >2.00 are considered good for urethane foam, whereas values <1.75 are considered poor.

Impeller term used to describe the power-driven mixing blades that are used to mix urethane chemicals in a mixing head.

Index in the case of urethane foam, it is the relationship between the equivalent weights of the isocyanate and the water and polyol. An index of 100 indicates both are balanced. 95% indicates a shortage of 5% isocyante, and 105% an excess of 5%.

Inhibitor a substance that slows down a chemical reaction. Sometimes used to prolong shelf-life.

Initiation time a synonym for cream time.

In-line cutters cutting machines installed directly on a continuous foaming line to save foam handling time.

In-line mixer a special mixer that has been added to pre-mix one or more minor components before entering the main component stream.

Instron an instrument used to determine the mechanical properties of foam.

Integral skin foam a moulded urethane foam product having a dense outer skin and soft core.

Interlocked system practice of using a single power source to drive two or more fluid metering systems.

Internal mix practice of mixing the various components inside a mixing vessel and then discharging the mixture through one or two ports.

Intumescence the foaming and swelling of a plastic when exposed to high surface heat/flames.

Irregular cells describes a foam having a mixture of widely varying cell sizes.

Isocyanate family name for those chemical components having one or more reactive nitrogen–carbon–oxygen radicals or groups attached to the main molecule.

Isomer any one of two or more chemical compounds having the same number of the same kind of atoms in their structure.

Isomer ratio is the ratio between the 2,4 isomer and 2,6 isomer in commercial toluene diisocyanate. The 80/20 blend is most commonly used, although the 65/35 isomer blend is also available.

Isotropic foam characterised by having the same strength properties in all directions.

Jacketed tank term used to describe a tank having an additional shell with cooling or heating devices.

K factor a measure of the insulation ability or thermal conductivity of a foam material.

Knit line a high-density layer of foam indicating the joint between two adjacent pours of foam that did not flow together until after skins were formed.

Lag term which refers to the delay or reduction in flow that occurs when a restriction in the flow causes a build-up in pressure.

Lamination process of adhering two or more sheets together to form a thicker product.

Lead lag describes the problem of non-uniform flows of components at start-up.

Lead time term sometimes used to describe the time between the end of mixing and pouring.

Linear molecule a long-chain molecule as contrasted to one having many side chains or branches.

Loose skin a condition in foam moulding in which the skin of the moulded product is a loose intact film.

Machine tanks tanks that form a part of a foam machine with metering pumps.

Manual flush a manual operation to clean the mixing head with solvent.

Mass effect term used to refer to the influence the total amount of chemicals poured has on the density and cure time.

Masterbatch describes a mixture resulting from premixing as many as possible of minor ingredients with polyol. Also refers to a concentrated colourant in a premix.

Matrix in the case of urethane foam, the bubbles originate in a liquid matrix.

Mattress a cushioning material on a bed to maximise comfort.

Mechanical mixing head a mixing device with power-driven rotor and impeller.

Mer a basic unit in a plastic and the repeating structural unit in a polymer.

Metering the calibrated flow of liquid chemical in a foam formulation.

Metering system the pumping system used to accurately control the flow of liquids.

Methylene chloride a chlorinated hydrocarbon commonly used as a secondary blowing agent and also used as a cheap cleaning solvent in the urethane industry.

Mill in the foam industry a 'hammer mill' is used to produce fine particles from foam scrap.

Mixer a mechanical device capable of mixing together two or more materials homogenously.

Modified pre-polymer a polymer system that has been modified by diluting the catalyst system with plasticisers and additives.

Modulus of elasticity the ratio of stress to strain in a material that is deformed elastically.

Mould liners thin films of plastic or treated paper used to line the inside of mould cavities and used instead for mould-release agents. This is a must in foam production because of foam adherence.

Mould packing small percentage of excess total liquids poured to counter weight loss because of gas.

Mould-release agents chemical compounds which, when applied to the insides of a mould or foam carrier, prevent the cured foam from sticking to the mould.

Moulding practice of pouring a foam liquid mix into a mould cavity to mould a shaped product.

Monitoring checking of foaming process at various stages of production and after.

Nominal throughput range term used by foam machine manufacturers to indicate the throughput or flow rate capacity that particular machine was designed to handle with normal formulations.

Nozzle very general term used to describe the discharge opening or discharge port of the mixing head.

Nucleation term used to describe the generation of small uniform bubbles by injection of air or other liquid chemicals to improve foam quality.

Open cells cell structure characterised by open interconnecting cells or bubbles. Best quality flexible foams are obtained by 100% opened cells.

Open moulding practice of pouring a foam mix into a large enclosed mould with an open top.

Organotin catalysts family of urethane foam catalysts that are used to help in the blowing and gelation reactions. A very common tin catalyst is stannous octoate.

Overfill amount of unwanted excess of foam liquids poured into a mould intentionally or otherwise.

Preflex practice of compressing a flexible polyurethane sample up to six times to a pre-determined thickness before determining the indentation force deflection.

Paper-handling device device to cradle a large reel of paper to unwind over a special framework, such as on a continuous foaming machine.

Peak rise point the point of maximum foam rise or height on a continuous foaming line.

Peelable quality refers to round foam blocks having a very uniform cell structure allowing peeling of very thin thicknesses (e.g., foam sheets for quilting on mattresses).

Peeling machine mechanical cutting device capable of slitting or peeling a continuous sheet of foam (usually for quilting and packing).

Permeability rate at which a liquid or gas can penetrate through a material.

Pillows foam pillows can be moulded or cut from foam blocks. A common practice is to also make pillows from small-particle foam waste after shredding.

Pin impeller rotating mixing blade with a straight shaft having a series of round-, square- or hexagonal-shaped 'pins' mounted at right angles to the shaft. There are also other variations.

Pin mixer a mixing head using a pin impeller.

Plastic one of many high polymeric substances, including natural and synthetic products but excluding rubbers.

Plasticiser various chemicals used as additives in a formulation to increase the flexibility of a foam or plastic product.

Plateau describes a flat spot on a foam.

Pneumatic drive better known as air-actuated motion drives (e.g., pneumatic cylinder).

Pocket blows tears or rips in the foam. Generally occur on the sides or inside a foam block.

Poise a measure of the specific viscosity (thickness) of a fluid.

Polycone a cone-shaped nozzle used to control the dimensions of the liquid stream coming from a mixing head.

Polyester polyol used for urethane foam production but the foam is rough and abrasive.

Polyether polyol used for urethane foam production to get soft, flexible foam.

Polyisocyanates isocyanate compounds having more than one isocyanate (nitrogen–carbon–oxygen) group attached to the molecule.

Polymer a high-molecular weight organic compound (natural or synthetic) whose structure is made up small repeating units (mers).

Polymeric isocyanate generally refers to those isocyanates containing more than two isocyanate groups in the molecule.

Polymerisation the chemical reaction by which larger molecules are formed by the joining of smaller molecules.

Polyol chemical compound with more than one reactive hydroxyl group attached to the molecule.

Polyol mix the product resulting from premixing several compatible minor ingredients with polyol.

Polyurethanes refers to the reaction of polyisocyanates (nitogen–carbon–oxygen) and polyhydroxyls (hydroxyl) groups. Complex polyurethanes are based on toluene diisocyanates and polyester and polyether polyol chains. Other isocyanates used are diphenylmethane and polymethylene polyphenyl. Polyurethane foam can be produced as a flexible, semi-rigid and rigid product of different densities. Flexible foams will have open cells, whereas rigid foams will have closed cells.

Pore diameter cell size.

Portable cutters small, lightweight cutting devices/machines for cutting foam.

Positive displacement refers to pumps that are so designed that substantially all of the material sucked is delivered through the outlet of the pump.

Positive metering the ability to control the flow rate of a particular fluid with accuracy to a great extent.

Post-cure period of curing time after the foam blocks have been removed from the machine or conveyor. During this time, heat is given out and post-cure for conventional flexible foam is ≥24 h and ≥48 h for viscoelastic foam. Very good ventilation is required during this period to counter possible fire hazards.

Pour-in-place practice of pouring a liquid mix into a cavity and having it filled with foam.

Pour pattern the pattern formed by the liquid stream being deposited in a mould cavity or conveyor; this is often quite critical.

Pouring head a mixing and dispensing head designed for chemical liquids.

Pouring point the position at which the mixed foam liquid is deposited on the conveyor of a continuous foaming system.

Pre-cure time interval between pouring the chemical mix into a mould and removing the product without damage. Sometimes called the 'handling time'.

Pre-cure oven an oven in which heat can be supplied for accelerated curing.

Powered conveyor a conveyor having the moving surface powered by a drive motor.

Pre-heat technique of heating moulds or cavities to desired temperatures before pouring.

Premix term often used synonymously with masterbatch or a polyol mix.

Premix tanks tanks used for making batches of premix, separate from the machine.

Pressure gauge device for measuring/checking pressure. All urethane foam-processing machines are equipped with accurate pressure gauges in all flow circuits for process safety.

Primary alcohol groups reactive groups present in certain polyol molecules. Usually the higher the percentage of primary alcohol groups, the less catalyst required for curing.

Process variables factors in chemical foaming process such as the machine, room temperatures and others that could affect the quality of the foam.

Pumping impeller mixing blade designed to accurately move the liquids being mixed through the mixing zone and through the discharge orifice with uniform force.

Recovery amount of return to original dimensions and properties of a flexible foam sample.

Resilience indicator of the surface elasticity or 'springiness' of a sample of flexible foam. It is measured by dropping a standard steel ball onto a foam cushion from a given height and measuring the percentage of the ball rebound.

Rat holes refers to the large, irregular, normally elongated gas pockets usually found in frothed foam.

Ratio control ability to change and regulate the ratio or proportion between two or more fluid components.

Raw materials term for chemical ingredients used for manufacturing foam.

Reactants general term for any of the raw material or intermediates used in the manufacture of foam.

Reaction balance the important balance between the relative reaction rates of gas generation (blowing) and gelation (polymerisation).

Re-bonding term used to describe the process of adhering small shredded particles of foam scrap into sheets, blocks, products as saleable items (e.g., carpet underlay).

Recirculation system refers to the practice of continuously pumping the metered fluids back to the machine tanks during the 'off' period of a production cycle. The 'on–off' valves in the mixing head serve to divert the continuously metered flow into the mixing chamber or back to the machine tanks.

Regulator device to control the flow of fluids.

Relaxation term used sometimes to describe the settling of foam, after rising and peaking.

Remote drive a drive motor usually for a mixing head that is located at some distance.

Reodorant a powerful pleasant-smelling chemical which is added to a formulation to mask unpleasant odours sometimes associated with amine catalysts or even some polyols.

Resilient foam foam that has very rapid recovery from extreme compression and a fairly high increase in linear resistance.

Resin term used to describe the unsaturated polymers or monomers. Also called plastic resins.

Rheology the study of low and deformation of matter.

Ribs thread-like structures formed at the joints between adjacent bubbles in a foam which become open cells.

Rigid foams a class of stiff closed celled foams which may be thermoforming or thermosetting.

Rise time time between the liquid mixture being poured and the completion of the foamed mass.

Run time time during which a foaming machine is actually in operation.

Slab stock flexible polyurethane foam made by the continuous pouring of mixed chemical liquid onto a moving conveyor and cutting into desired lengths using a cut-off saw mounted on the conveyor.

Static fatigue loss in loadbearing properties of a flexible foam sample after being under constant compression.

Support factor the ratio of 65% indentation force deflection to 25% indentation force deflection; a compression test. A support factor >2.0 is considered to be good.

Surfactants chemical compounds used in a foam formulation to help and control cell formation during a foaming process.

Schematic diagram drawing that illustrates the design and layout of a machine, circuit or process in an abstract or symbolic manner.

Scoop cutting technique for producing special cut shapes from sheets or blocks of flexible foam by cutting an impression on foam by compression.

Scrap generally refers to waste from urethane foam manufacturing process. An inherent waste is generated from the top, side and bottom skins of foam blocks.

Seat cushions blocks of foam of different shapes and densities selected for particular applications. Sitting cushions have a higher density than back cushions.

Secondary alcohol groups reactive alcohol groups present in most polyol molecules.

Self-extinguishing ability of a foam to stop burning after it has started burning.

Settling normal loss in height of foam after peak rise.

Shaping saws cutting devices that can produce any special shape desired from a foam block or sheet.

Shear resistance ability of a flexible foam to resist laterally applied forces.

Shiny foam foams with a high proportion of uniform cell membranes that glitter from reflected light.

Shot total amount of mixed chemical liquids dispensed from a mixer during one cycle.

Shot cycle describes the total time spent on 'on' and 'off' in a single unit of operation.

Shot cycle timer a timing device used for automatic control of a shot cycle.

Shredder a mechanical device used to reduce foam waste to small pieces.

Shrinking describes the loss in size that occurs during foam production. Can occur because of excess heat or faulty formulation.

Side cracks definite cracks in foam blocks or slab stock with widely separated edges.

Side splits horizontal tears or rips in the side of foam blocks. These splits are horizontal to the plane of the foam conveyor or slightly angled.

Silicones complex chemicals formed from a combination of silicon and organic groups. These are widely used in urethane foam formulations for cell structures.

Skins describes the high-density outer surface of a foam block.

Slab foam foam made by the continuous pouring of mixed chemical liquids on a moving conveyor, which generates a continuous loaf of foam.

Solid elastomer rubber-like compounds that have no internal cavities or gas bubbles.

Solid filler an insoluble additive used in urethane foam formulations. Adds weight and reduces the cost of foam products.

Solvent a substance (usually liquid) used to dissolve another substance.

Solvent blown foam foam made with low boiling chemical compounds as blowing agents.

Solvent flush refers to cleaning of mixing tanks, chambers or mixing heads after each production cycle. Done manually or automatically.

Solvent flush timers timers used to control automatic flushing systems.

Specific gravity the density (mass per unit volume) of any material divided by that of water.

Splashing a splattering that occurs if the mixed liquid is poured into a mould or onto a moving conveyor.

Sponge flexible or semi-rigid foam with irregular small-to-large cells that helps absorption and easy let out of liquid if squeezed.

Spreader device to assist in placing a uniformly distributed layer of mixed foam liquids.

Stabilisers additives such as antioxidants and ultraviolet absorbers that help to maintain the quality of foam in use. Silicone stabilisers add stability and control to the semi-liquid foaming mass.

Stannous octoate one of the commonly used organotin catalysts used in flexible foam manufacture.

Stannous oleate organotin catalyst less active than stannous octoate.

Static fatigue describes the loss in the loadbearing properties of a foam.

Stator blades non-rotating or non-moving baffles in a mixing chamber that assist mixing foam components.

Stoichiometry refers to the relationship of the various combining weights of the several interacting chemicals.

Stream term for describing the flow of liquid of one of the components in a formula.

Streamers lines or streaks of cream-coloured liquid extending upstream from the cream line into the clear liquid dispensed from the mixer during continuous foaming.

Surface-active agent an additive to a foam formulation that helps or hinders the formation of a fine, uniform cell structure in the resulting foam.

Synergism property exhibited by a blend of some materials having a greater effectiveness or chemical activity as a mixture than would ordinarily be expected from the sum of their independent abilities.

System rather ambiguous term used to describe almost any combination or method of mechanical parts or chemicals that have some relationship to each other.

Tear strength a measure of force required to continue to tear a foam sample after it has to tear.

Tensile strength a measure of the force required to stretch a standard sample of material to its breaking point.

Tight foam polyurethane foam with many closed cells, resulting in low air flow.

Tachometer a device to indicate the speed of some mechanism (usually the number of revolutions per minute).

Tack-free time time between pouring the liquid mixture and the time that the foam surface can be touched without sticking.

Tacky surface the condition of the surface of a foam block which is still sticky.
This could be due to insufficient cure time or faulty weighing or faulty formulating.

Tear resistance ability of a foam sample to resist tearing after a tear has been made.

Tertiary amine an amine catalyst used in urethane foaming. Tertiary amines are more powerful than primary or secondary types of amines.

Thermal conductivity ability of a material to conduct heat. Measured by the quantity of heat that passes through a unit cube of substance in unit time if the difference between the two faces is 1°.

Thermal gravimetric analysis analysis of weight loss of a foam sample at different temperatures through the decomposition point or zone.

Thermoplastic a material which is capable of softening or melting at elevated temperatures without degradation and on cooling retains its original condition. Material can be re-used, for example, polyethylene and polystyrene plastics.

Thermosetting a material that is fully cured by elevated temperatures into a solid condition and cannot be recovered by re-heating. Material cannot be used again, for example, polyurethanes.

Thixotropic refers to the ability of a fluid to be semi-solid at rest but reverts to a liquid upon being agitated or stirred.

Three-roll paper system a system of paper supply for a conveyor consisting of a single bottom sheet of paper with two separate side sheets. Required for a continuous foaming system.

Throughput control being able to vary the flow rate of a metering machine.

Throughput indicators any one of the several devices that have been calibrated for flow rates such as tachometers and flow meters.

Throughput rating applies to metering machines and mixing heads, and indicates the maximum flow rate that can be metered.

Tolerance a specified allowance for deviations in measurements.

Top cracks cracks that occur on the top surface of a foam slab that extend across the width of the conveyor.

Top cure process of shortening the tack-free time of foam blocks or slabs by heat.

Transfer pumping refers to the method of filling the foam machine tanks or keeping them filled.

Trichlorofluoromethane a chemical name for a fluorocarbon. A blowing agent or auxiliary blowing agent for urethane foam.

Trim resulting foam from trimming crude foam blocks or slabs before converting to products. After shredding, these can be converted into saleable products.

Triol a polyol characterised by having three reactive hydroxyl groups attached to the molecule.

UV stabiliser chemical compounds used to counter degradation by ultraviolet light.

Ultraviolet zone of invisible radiations beyond the violet end of the spectrum.

Undercutting a condition that can result in flow lines, side and top cracking as well as the entrapment of large bubbles in the centre of the foam slab.

Underfill describes the condition of insufficient pour of chemical mix resulting in not filling a mould fully.

Urethane the product resulting in a chemical reaction between a chemical containing reactive isocyanate groups on its molecule and a chemical containing reactive hydroxyl groups on its molecule. Urethane is the basic 'family' and polyurethane is one of them.

Velocity balance a lead-lag problem under some conditions of metering and mixing machine operation in which the pressures are balanced exactly.

Vertical cutting saws special foam-cutting bandsaws in which the blades used for cutting are in a vertical plane, and they cut the foam block from top to bottom.

Vertical pour term generally refers to pour-in-place operations.

Virgin foam refers to flexible slab or block foam that has not being processed in any manner other than trimming all sides and the bottom.

Viscoelastic refers to a foam having four-dimensional properties such as density, temperature, time effects and hardness (unlike standard conventional foams).

Viscous describes a flow condition of any fluid more resistant to flow than water.

Water absorption degree of absorbing water when foam is immersed under certain conditions.

Water-blown foam foam made by expansion of gas generated by the reaction between water and an isocyanate.

Windows cell membranes or walls in flexible foam that are unbroken or ruptured but intact, which interfere with free air movement through the foam structure.

Yield point the point at which in compression or tension loading of a foam sample beyond which permanent deformation occurs.

Yield strength the force required to bring a specified sample of foam to the yield point.

Index

https://doi.org/10.1515/9783110643183-016

www.ingramcontent.com/pod-product-compliance
Lightning Source LLC
Chambersburg PA
CBHW061417210326
41598CB00035B/6243